# Beyond Walden

# Beyond Walden

The HIDDEN HISTORY *of*
AMERICA'S KETTLE LAKES *and* PONDS

## Robert M. Thorson

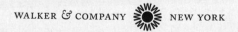

WALKER *&* COMPANY    NEW YORK

All photos and images are courtesy of the author, with the
exception of the following: p. 5, courtesy of the Walden
Woods Project; p. 48, used by permission of R. Dale Guthrie;
p. 69 (MHS Image ID 1644) and p. 100 (MHS Image ID 60713), used
by permission of the Minnesota Historical Society; p. 115 (WHS
Image ID 36627), used by permission of the Wisconsin Historical
Society; and p. 173, adapted by the author from an illustration by Guy
Troughton in *The Natural History of Lakes*, by Mary J. Burgis and Pat
Morris (Cambridge, UK: Cambridge University Press, 1987).

Published by Walker Publishing Company, Inc., New York

LIBRARY OF CONGRESS CATALOGING-IN-PUBLICATION DATA

Thorson, Robert M., 1951–
   Beyond Walden : the hidden history of America's kettle lakes
and ponds / Robert M. Thorson.—1st U.S. ed.
        p. cm.
   Includes bibliographical references.
   ISBN-13: 978-0-8027-1645-3 (hardcover)
   ISBN-10: 0-8027-1645-8 (hardcover)
   1. Kettle holes—United States—History.   I. Title.
   GB621.T48 2009
   551.48'20973—dc22

                                                        2008041839

Visit Walker & Company's Web site at www.walkerbooks.com

First U.S. edition 2009

1 3 5 7 9 10 8 6 4 2

Book design by Simon M. Sullivan
Typeset by Westchester Book Group
Printed in the United States of America by
Quebecor World Fairfield

*To my twin brother, Jim, who probably likes kettle lakes*
*as much as I do, but for quite a different reason.*

A lake is the landscape's most beautiful and
expressive feature. It is earth's eye.

HENRY DAVID THOREAU, *Walden*


This was our very own lake, filling our life to the brim.

STERLING NORTH, *Rascal*

# Contents

*Lake Plantagenet, Minnesota.*

# Introduction

ON NOVEMBER 15, 1620, Captain Myles Standish and a party of sixteen well-armed Pilgrims made their first foray deep into the woods of America. Their route took them away from a sandy beach on Cape Cod Bay and across a narrow neck of land known today as Truro, Massachusetts. Though they failed to encounter the Indians they were pursuing, they did succeed in becoming the first English settlers to experience the joy of America's kettle lakes, which are called ponds in New England.[1]

"Kettle" was the word used by the Old Norse, the Goths, and the Anglo-Saxons for a rounded cooking vessel, typically made of iron: spelled *ketill, katil,* and *ketel,* respectively.[2] The word "lake" comes from the Teutonic root *lac,* denoting moisture, as applied to pond, bog, and stream. More than a millennium later, the words were combined to describe a special type of glacial lake created by downward or inward melting of a stagnant block of ice:[3] in simple terms, a meltdown depression filled with freshwater; even more simply, a "fossil iceberg."[4]

After an arduous journey and becoming "sore athirst," the Pilgrims found "springs of fresh water" draining from a kettle at higher elevation, from which they drank their "first New England water with as much delight as ever we drunk drink in all our lives."[5] Later the party discovered and excavated a storage mound full of dried corn built by the Indians, unearthing a "great kettle . . . brought out of Europe." They also discovered "a fine clear pond of fresh water, being about a musket shot broad" and surrounded by "much plain ground, about fifty acres, fit for the plow." From this beautiful blue glacial kettle they dipped water using the manufactured metal kettle. After reboarding the *Mayflower,* they crossed Cape Cod Bay to found Plymouth, the first successful European colony in the northern United States. Their journey took them across what had once been a muddy glacial lake, and before that, a massive lobe of slowly creeping ice.[6]

The ponds encountered by the Pilgrims were indeed kettle shaped, having smooth, circular outlines and steep shores, and resembling football stadiums built below ground level and filled with water. There the meltdown was local, as if a trapdoor in the sand had been opened from below. Much bigger kettles define the northern lake country of Michigan, Wisconsin, and Minnesota, hereafter called the heartland. These have ragged contours, with marshy bays, boulder-studded peninsulas, and gently curved beaches. They formed when multiple slabs of ice the size of villages became stranded during glacial retreat and were too big to be buried completely. There the meltdown collapse was inward, as if an icy retaining wall had been pulled back from a sandbank.

The vast majority of kettle lakes, however, are neither small and circular nor large and jagged. Instead they are medium-sized depressions, ranging from a quarter mile to two miles across, and having only slightly irregular outlines.[7] These inauspicious bodies of water are the most common and widely distributed "species" of natural lake in the conterminous United States.[8] Unlike normal lakes, which usually have significant inlet or outlet streams, kettle lakes are natural wells tapping the groundwater table, their waters filtered through silica-rich, grit-free sand. Being isolated and randomly distributed on the landscape, each lake and pond can have its own personality and its own unique physical and chemical balance between too much and too little heat and oxygen and too many and too few nutrients and dissolved salts. They are highly sensitive to pollution, vulnerable to climate warming, and individually quirky.

The natural resources of kettles have played a crucial role in American cultural history ever since nomadic ice-age hunters butchered mastodons on their shores about eleven thousand years ago.[9] Wild rice grown in the marshes bordering kettle lakes became the "staff of life" for northern Algonquin tribes, especially the Ojibwe, and fish became their mainstay protein. Beaver pelts obtained from the shores of countless small lakes became the coin of the realm during the European fur trade, which began in the mid-sixteenth century with Jacques Cartier. Colonial farm villages of the seventeenth and eighteenth centuries depended on kettles for hay meadows and hydropower, respectively. They also provided bog ore, ice for refrigeration, and cranberries, which prevented early American sailors from getting scurvy.

Kettles continued to influence the American settlement during the era of Manifest Destiny. As New Englanders leapfrogged west during the nineteenth century, they settled in terrain where kettles were common—north of the glacial limit but south of the harder rocks of Canada. Culturally, this lay above areas settled by a polyglot of Germans, Dutch, and Scotch-Irish who moved west via the Cumberland Gap, but below those of French Canadians who moved west via the St. Lawrence Seaway and the upper Great Lakes. As a result, the architectural styles of early houses from Maine to Minnesota follow the "English Tradition," dominated by Yankee, saltbox, and Greek revival motifs. Before statehood, the Minnesota Territory thought of itself as a "New England West." Politically the kettle lake band tends to be socially conservative, intellectually liberal, and particularly suspicious of Southern ideas. Not one kettle lake state joined the Confederacy.

After the Civil War, to borrow a phrase from historian John Lewis Gaddis, the band of kettle lakes continued as a "landscape of history."[10] Near the edge of the prairie, their beauty was used as bait to lure immigrants from Scandinavia and elsewhere in northern Europe where Old World kettles had been part of their homeland. The supply of potable water within them helped draw immigrants out onto the prairie and nucleate their homesteads.[11] As America entered the twentieth century, however, the densely forested, lake-speckled interiors of Michigan, Wisconsin, and Minnesota were still islands of quasi wilderness surrounded by settled lands. This largely uninhabited terrain would soon become a recreational paradise, a middle-class utopia of summer lakeshore homes.

. . .

AMERICA'S MOST FAMOUS KETTLE is Walden Pond, in Concord, Massachusetts. During the 1840s, its shimmering northern shore was owned by the poet-sage Ralph Waldo Emerson, who doubled as an American country squire. It was the crown jewel of his "lake district," modeled after that of the English romantic poets, especially William Wordsworth and Samuel Taylor Coleridge. The inspiration Emerson drew from walking Walden's woods helped steer America away from the spiritual ice that was New England Puritanism toward the natural warmth of transcendentalism, and helped make Boston the "Athens of America." Water from nearby Fresh Pond, a kettle in Cambridge, still nourishes two of the area's great universities, Harvard and MIT.

But it was Emerson's protégé, a young bachelor named Henry David Thoreau, who made Walden famous. He went beyond using the pond to inspire poetic thoughts. He actually fell in love with the place. The potency of his affection translated this perfectly ordinary, isolated, cobble-rimmed kettle into great, but enigmatic, literature. Where his daytime eyes saw cranberries on rounded tussocks of peat moss, his mind saw crimson "jewels worn or set in those sphagnous breasts of the swamp." Where his nighttime eyes saw the "soft and velvety light" of the aurora borealis reflected on the pond, his mind saw "placid days . . . put to rest in the bosom of the water."[12]

Thoreau's *Walden*, published in 1854, gave rise to a uniquely American way of thinking about nature that erupted a century later into a political movement called environmentalism. Thoreau admitted that "the scenery of Walden is on a humble scale, and, though very beautiful, does not approach to grandeur."[13] But it was this ordinariness that allowed Walden Pond to stand for every small lake in America, every rural retreat, and every intimate place where something natural was being threatened by the human stain. Rachel Carson, whose 1962 *Silent Spring* is considered the tipping point away from the cold war arrogance of "better things through chemistry" toward the sensibility of better things through ecology, "kept a well-used copy of *Walden* by her bedside," as if it were a Bible. "No parallel tract or body of water or place," the historian W. Barksdale Maynard writes, "has so

*Henry David Thoreau, daguerreotype by Benjamin D. Maxham, 1856.*

captivated the human imagination." Walden has become "an international shrine."[14]

During the middle of the twentieth century a completely different version of kettle lake appreciation emerged in the northern heartland states of the upper Midwest, as kettles became pastoral retreats for middle-class families. There when folks say they're heading "up to the lake," their plan is not to sit alone on a pine stump and meditate on ultimate meanings. Their plan is to drive north and go fishing, enjoy a breezy sunset, have a family picnic, take a boat ride, or let the children chase frogs. The goal is family fun and relaxation in a setting midway between wilderness rigor and city bustle. Garrison Keillor poignantly captured this sentiment in a body of water called Lake Wobegon, "blue green and brightly sparkling in the brassy summer sun and neighbored by the warm-colored marsh." As with Walden Pond in an earlier era, the simple beauty of Lake Wobegon elevates the mundane town near its shore. By keeping the location secret—somewhere in "tiny Mist County," the "phantom county in the heart of the heartland"—Keillor's creation became the fictional archetype for every perfectly ordinary kettle lake or pond in the country.[15] Everyone's favorite lake can be imagined a personal Lake Wobegon, symbolizing the disconnect between life as it is—the alienation of factory, office, and freeway—and life as it should be—a community of nature-loving people enjoying themselves amid crystalline lakes and scented forests. Keillor put the northern heartland's mélange of kettle lakes on America's cultural map, much the way Mark Twain did the Mississippi River, James Fenimore Cooper New York's Leatherstocking Region, Willa Cather the prairie, Mary Austin the desert, and John Muir the Sierra Nevada.

THOUGH SEPARATED BY more than fifteen hundred miles, kettles near Boston and Minneapolis have much in common. They are local clusters within a single galaxy of small lakes that sweeps across nineteen northern states between the Atlantic Ocean and the Rocky Mountains. It extends from the foggy, boggy ponds of Nantucket Island, Massachusetts, to the sunbaked prairie potholes beyond Great Falls, Montana. As with the shape of the Milky Way, the lake galaxy is thickest near the middle, between Michigan's Upper Peninsula and central Indiana, thinning eastward and westward. On a satellite view of the region taken during the summer, one could easily count over

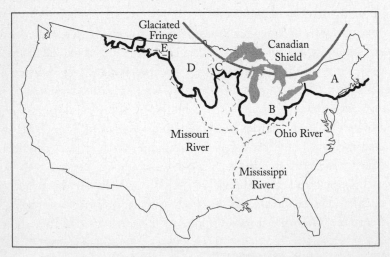

*The sectors of the glaciated fringe are A) Northeast, B) Great Lakes,*
*C) Superior, D) Dakota, and E) Montana.*

fifty thousand blue dots on a backdrop of green woodland and prairie. In winter they become snow-white freckles on a grainy, sepia-toned backdrop of dormant trees, windblown grasslands, and the lines of the human-built environment.

Each kettle is interesting in some way. Salt Pond in Eastham, Massachusetts, may be the most overlooked one in the nation: Millions of visitors look beyond it to catch their first glimpse of Cape Cod National Seashore. Jamaica Pond is the centerpiece of Boston's Emerald Necklace, a pond-themed chain of parks designed by Frederick Law Olmsted, the founder of landscape architecture in North America. Long Island's Lake Ronkonkoma resembles a bright blue meteorite-impact crater smack-dab in the suburbs. Pennsylvania's Lake LeBoeuf was visited by George Washington when he was a young soldier in British uniform during the run-up to the French and Indian War. Maine's Schoodic Lake was visited by Jefferson Davis— soon to be president of the Confederacy—just before the Civil War (some thought he was on a spy mission). Boot Lake in Elkhart, Indiana, is arguably America's most volatile kettle, having gone from swimming pond to illegal dump to sludge farm to environmental education center within the span of a century. John Muir, America's first heroic preservationist, almost drowned at Fountain Lake, Wisconsin, where he grew up. Lake Itasca, Minnesota, is the spring-fed source of the Mississippi River. Kettle Lake, North Dakota, was an important oasis for migrating waterfowl six thousand years ago, when the climate was drier, perhaps an omen of things to come in our steadily warming world.[16]

Geographically, kettles occur within the glaciated fringe, a nearly continuous but highly irregular arc of sandy glacial deposits lying between the clay-rich soils south of the last major ice advance and the rocky terrain to the north in Canada. When Walt Whitman looked at the fringe, he saw an east-west belt of great oval lakes—Superior, Michigan, Huron, Erie, and Ontario—between Minnesota and New York. What I see is a continuous band of lakes from Montana to Maine, in which sandy kettles are sprinkled like confetti between these rock-carved inland seas.

The distribution of kettle lakes and ponds within the United States breaks down into five natural regions.[17] Southernmost New England contains a sandy archipelago extending from Nantucket to New York City, including Cape Cod, Martha's Vineyard, the Elizabeth Islands, and Block Island, Fishers Island, Long Island, and Staten Island. Practically every natural body of freshwater in this region is a kettle lake or pond. Most formed in association with ridges of sand and gravel called moraines, thrust up by the ice sheet near their outer limits.

The second region sprawls across the interior of New England and upstate New York. There kettles occur in clusters on patches of sand and gravel, and within a wooded rural-suburban landscape of stony hillsides, smooth streamlined ridges called drumlins, and subdued cliffs of rock called ledge. Equally common are partial kettles, a term referring to those lakes formed by sediment burial and collapse along only part of an ice-block perimeter.[18]

The third region approximates the divide between the St. Lawrence and Ohio rivers, from western New York to northeastern Illinois; its kettles occur as a discontinuous sprinkle of small lakes hiding in otherwise productive and well-drained agricultural landscapes.

The fourth region includes the treeless High Plains, where kettles are the local standouts within an enormous population of prairie potholes, small dots of marsh and open water within an ocean of green and tawny grass.

But it is in the fifth region—the northern heartland states of Michigan, Wisconsin, and Minnesota—that kettles are signature landforms.[19] Only there do they dominate and define the landscape, as do the red-rock canyons in the American Southwest, the barrier islands on the Atlantic coastal plain, the caverns in the central Appalachians, and the volcanoes surmounting the mountain ranges of the Pacific Northwest. These were the lakes that I learned to love as a child, having been born in Wisconsin, summered in Minnesota, where my family later moved, and also lived in North Dakota and Illinois.

The study of kettles helped shape American science. For several decades G. Evelyn Hutchinson, professor of zoology at Yale University and the father of American limnology, the science of lakes, used a small kettle in Branford, Connecticut, named Linsley Pond as a veritable outdoor petri dish to teach students about the interactions between different species, water chemistry, and solar energy.[20] His work helped launch ecology as a quantitative science distinct from its qualitative predecessors zoology and botany. Land-grant campuses in Wisconsin, Michigan, and Minnesota pioneered a broader approach to lake science at watershed scales. Kettles have also contributed greatly to our understanding of watershed soils, groundwater hydrology, and climatology. Nowhere at midlatitudes is there a better fossil record of how ecosystems have responded to climate change.

MANY SMALL LAKES are imperiled.[21] Heated debates take place over whether Jet Skis should be allowed, whether lakes should be treated with chemicals to control weed growth, and whether trophy fish should be stocked to raise revenues from fishing licenses. In each case the eternal struggle between private freedom and the public good is being magnified by the struggle between recreational development and conservation. Traditional lakeshore residents don't often think of themselves as being responsible for the degradation of lake quality. Yet they automatically become part of the problem whenever they flush a toilet, fertilize their petunias, feed the dog, take a boat ride, or dump a minnow bucket without paying attention to what's inside of it first.

As a group, kettles are slightly healthier than other types of lakes of comparable size. Those that remain undeveloped are almost as vibrant as they were a century ago, although they are being invisibly damaged by aerosol fallout from remote sources of air pollution. During the 1970s kettles with small cities and wastewater treatment plants on their shores were seriously degraded, but they have improved. Now the main threat facing kettles is too much development, both of their shorelines and of the highly permeable soils between them. Excess nutrients from covert sources like septic tanks, feedlots, and tillage fields contribute to weedy overgrowths, murky water, and fish kills. Other problems include the closure of public swimming beaches due to fecal bacteria, the ecological leveling caused by invasive species, pharmaceutical pollution, and bioaccumulated toxins in game fish. Future worries include overdevelopment, the long-term effects of climate change, and demographic shifts that are imperiling conservation and appropriate use.[22]

Modern lake management is mostly people management. Problems persist not because aquatic scientists and engineers don't know what to do, but because they lack the background data needed to make a diagnosis, the money to fill a prescription, and the legal power to implement rehabilitation. Future lake management will depend more on the efforts of private lake associations than on top-down government regulation to keep small lakes clean and to restore those that need help. *Beyond Walden* is written for all who cherish small lakes as precious sources of freshwater, relaxation, recreation, and natural habitat for plants and animals, and who want to help preserve them for the future.

## I.

# Ice-Sheet Invasion

THE KETTLE LAKES AND PONDS of the United States are landforms of transition. Geographically they preside over the sandy and gravelly soils between more loamy unglaciated terrain to the south and the hard-rock, ice-scoured terrain of Canada. Chronologically they were born during the last climatic flip-flop between glacial and interglacial modes. Ecologically a white blanket of ice became a biologically rich landscape of sky blue waters, green forests, and russet brown marshes and bogs. Between these scenes was a chaotic meltdown featuring a palette of textured grays.

This dramatic environmental transformation was triggered by an astronomical wobble. A toy top spinning at medium speed on the floor wobbles in a perfect circle around a stationary point on the ceiling. Earth wobbles the same way, but at a much slower pace, and with an axis of spin tipped toward a fixed celestial pole, the North Star. Though the total amount of sunlight on Earth remains the same during each wobble, the slight difference in the seasonal distribution of radiation is enough to nudge Earth's climate system back and forth between glacial and interglacial modes.

About twenty thousand years ago, the ice-making machine of the Canadian hinterland had wobbled to its full-on position. The amount of sunlight was more evenly divided between the seasons. Bitterly cold winters were slightly brighter, making them slightly warmer. Counterintuitively, this increased the snowfall, as with Antarctica in the warming world today.[1] Meanwhile, chilly summers became slightly dimmer and therefore slightly cooler. This reduced the snowmelt,

*A kettle forming in Denali National Park in Alaska.*

allowing more of the extra dose of winter snow to remain. Given this solar diet, the ice sheet began putting on girth and weight, eventually growing into a colossus nearly two thousand miles across and two miles high. Geologists call it the Laurentide Ice Sheet, named after the Laurentian Mountains of Quebec, which were completely buried.[2] At the Laurentide culmination, its southern edge traced an irregular arc from the Gulf of Maine to Montana, reaching its most southerly tangent at the latitude of Kentucky.

The mound of ice projected so far up into the atmosphere that it split the jet stream in half, diverting Arctic air as far south as Arkansas.[3] Lands bordering the southern edge of the ice were carpeted with tundra, a ground-hugging mixture of sedge, grass, heather, dwarf birch, willow, and alder, often whiskered by spindly spruce trees. Parsnip-shaped wedges of black ice grew on the flatlands of New Jersey, as do their modern counterparts in the Arctic. Patches of frost-heaved rubble called felsenmeer—German for "sea of rock"— formed in the Appalachians, especially in the highlands of Pennsylvania. During summers, a chilled layer of dense air flowed down the smooth surface of the ice sheet, compressing and drying as it descended. These katabatic winds sandblasted boulders and cobbles into curiously shaped stones called ventifacts that were polished to a high gloss. Billowing sheets of sand accumulated where the winds abated, forming dunes. A dust blanket called loess fell farther downwind, generally toward the east.

The water used to make the ice sheet came from the sea. As a result, the sea level dropped. Rivers of the coastal plain between New York City and Florida flowed more than a hundred miles across the now-submerged continental shelf. These east-flowing streams built sandy deltas that would later be reworked into barrier islands as the edge of the sea rose toward the west. The famous Outer Banks of the Carolinas were one result. Florida was half the size of Texas, just as dry, and choking with lime dust. Central Louisiana overlooked a Mississippi River that bore little resemblance to the meandering Big Muddy we know today. It was then a braided sand-gravel stream flowing within a shallow but enormous box canyon. In Maine the weight of the ice sheet pressed the land downward beneath the level of the fallen sea. There cliffs of floating glacier ice launched bergs into brackish water as fjord glaciers in Greenland do today. Southern New England was a bleak terrestrial landscape more than a hundred miles from the nearest Atlantic shore.

In the American Southwest what are now bone-dry mountains of sage and cactus were forested with aspen and pine. Parched desert basins became grassy green, or filled to overflowing with freshwater runoff. Modern Great Salt Lake is but a tiny, briny remnant of its ice-age predecessor, Lake Bonneville. Soaking rains on the southern High Plains from Texas to South Dakota were recharging aquifers, underground fractured rock or granular materials from which fossil water is being pumped dry today. Nebraska's Sand Hills accumulated during the geographic transition between dry-cold to the northeast and warm-moist to the southwest.

Ten thousand years later, Canada's ice-making machine had

wobbled to its full-off position. Northern hemisphere summers were slightly sunnier, winters slightly dimmer. Because one warm summer's worth of net melting easily makes up for several winters of net gain, the ice sheet had been put on a starvation diet. Ice had melted back well into Canada, its domes less than half their original height. The sea had reflooded much of its continental shelf. Like a chameleon, the Colorado Plateau had changed its color from glacial-age green to dusky earth tones. Meanwhile, an entire galaxy of kettle lakes had come into being.

SURPRISINGLY, THE BLUE WATER within kettle lakes and ponds is nearly twice as old as the rock within their shoreline boulders. Four billion years ago, Earth roiled with radioactive heat.[4] Its primordial atmosphere was toxic, filled with steam boiled out of the molten planet. This was the Hadean eon, named for the hellish conditions that dominated the first half-billion years of planetary history. Then came Earth's twilight eon, the Archean. Earth cooled to a temperature

*A Cape Cod kettle in Brewster, Massachusetts.*

at which a glaze of volcanic crust could form. Its steam atmosphere condensed, falling as a seemingly endless rain that created a universal ocean from the dark sky. That same water still cycles through lakes and ponds today.

Earth's heat engine continued to run at nearly full throttle. Miniature tectonic plates converged and diverged, creating an archipelago of smoldering volcanic islands like those of Indonesia today, but devoid of visible life. Beaches were jet black, with tinges of olive green. Ever so gradually, light-colored masses of more silica-rich rock—granite, gneiss, schist, and a special type of rock called greenstone—were smelted out of the darker crust to create masses of continental crust that would forever stand above the sea. This was the raw material that would later be broken, crushed, and rinsed to make the sand surrounding most kettles. During the subsequent Proterozoic eon, seven of these continental masses were welded together to create the stable core of the North American continent, its craton.

To the north these ancient basement rocks are exposed on the Canadian Shield, which is everywhere more than a billion years old. To the south they are mantled by younger, softer, and less-deformed marine sedimentary rocks of the platform, dating to Earth's present eon, the Phanerozoic, which translates as "visible life." To the west on the High Plains the platform is capped by much younger dinosaur-bearing shale. In the southern Great Lakes it is capped by middle-aged sandstone and limestone. New England is far too young to be part of the craton—its mountains still stand, albeit as the stubble of former summits. The raw material for most of its equally tough boulders was made less than half a billion years ago in the semi-molten root of ancient Appalachia, which towered above North America at a time when western mountains didn't exist.

When I was a teenager, I worked as a farmhand in North Dakota where my Norwegian ancestors had homesteaded. Using an old Ford tractor, chains, and a wooden sled, I plucked tough granite boulders from fallow fields and skidded them to growing piles started by my great-grandfather. Little did I know that they had come from the shield's Superior Province, which dates to about 2.5 billion years ago, making them more than a hundred thousand times older than the ice sheet that brought them there. Boulders in central Michigan and western New York came from a slightly younger though more exotic and far-traveled geological terrane called the Penokean Province. Those in the Finger Lakes of New York came from Grenville rocks, dating to barely a billion years ago. From North Dakota to New York, all of the boulders originated as jagged fragments torn from the cooled roots of an ancient mountain system before being milled into the shapes of potatoes as they were dragged away from the shield and scattered over the platform.

Some boulders are very special. Lieutenant James Allen of the Schoolcraft expedition to the Mississippi headwaters in 1832 reported one composed of pure copper, "bright on the surface, from the washing and abrasion of sand during freshets" and the "largest mass of native copper ever found." He estimated the boulder to be twenty cubic feet in volume and up to five tons in weight. Natives of the Old Copper Culture, dating to about five thousand years ago, mined copper from such boulders by breaking off chunks. These were then heated and worked into beautiful and useful objects such as pendants, spear points, fishhooks, axes, coins, and needles, which were widely traded from the Dakotas to the Atlantic, apparently at a time when the use of copper in the Old World was nascent. When Jacques Cartier encountered the natives near Montreal in 1535, he was told of the copper riches bordering Lake Superior, where more than five thousand prehistoric mining pits have been discovered, some up to eighty feet deep.[5]

Early natives also found strange, barn red and greasy gray, candy-striped boulders of iron ore. The red bands are hematite, precipitated from iron-rich brines on the floor of the sea when it was temporarily oxygenated. The grays are flint, precipitated in the sea under oxygen-poor conditions. Most of the iron and steel ever forged worldwide came from these ancient oceanic ores, especially those from the Mesabi Range of northeastern Minnesota.[6] Its ore was transported by rail to nearby Duluth, arguably the greatest iron mining port of all time. Owing to its heavy ship traffic on sea-sized Lake Superior, the Native American author William Least Heat Moon described it as "an old maid city looking under the bed each night for an ocean."[7]


SHAPING THE BOULDERS and moving them into position required a powerful ice sheet. Creating that glacial mass was the next step in making kettles. Two hundred and fifty million years ago, all of Earth's continents were assembled into a single supercontinent called Pangaea. Great tectonic rifts then sundered the land back into cratons. The fragment beneath what is now central Canada started out near the equator in a subtropical monsoon climate dominated by lumbering reptiles and oversized amphibians. It drifted northwest and rotated counterclockwise as the dinosaurs rose to prominence and went extinct. By forty million years ago, when the mammals were beginning their ascendancy, it had reached a latitude high enough to support seasonal snow. Meanwhile, the Antarctic craton drifted toward the South Pole until an ice sheet formed, ushering Earth into a period of cooling that's still with us today.

Within the last ten million years, Greenland became a carapace of ice, chilling the North Atlantic. Canada, Alaska, Siberia, Scandinavia, and Greenland migrated toward each other until they bottled up the Arctic, steepening Earth's north-south thermal gradient, and thereby increasing the flux of moist equatorial air toward the north. About two million years ago, the Isthmus of Panama shoaled, creating the Gulf Stream, which shunted warm water to higher latitudes. At last the North Atlantic had a vigorous atmospheric circulation capable of nourishing a large ice sheet on the craton of east-central Canada.

Ten ice sheets have come and gone within the last million years. Last was the Laurentide, born in Labrador on the final day of some subarctic summer about 110,000 years ago. Scientists have reconstructed its birthplace from the alignment of glacial scratches, subglacial rivers, drumlins, and strings of boulders traced back to their exotic sources. Like the rays of a circle, these lines have a common origin, a windswept tundra plateau in northern Quebec near the Labrador border between the fjord-serrated Atlantic coast and south-eastern Hudson Bay. The day of the Laurentide's birth was determined from statistical analyses of modern meteorological records, combined with studies of glacial mass balance. The year of its birth was obtained from laboratory measurements of radioactive and cosmogenic isotopes, especially argon and beryllium, respectively.

In the beginning, only a few patches of snow managed to survive the melt season. Because they reflected more sunlight than the adjacent drab green tundra, the air became locally chilled above the snow. Some of the precipitation falling as cold rain, which could drain to the sea, instead fell as snow, which could not. Late-summer melting diminished. Each year's accumulation of snow got a progressively bigger head start, resulting in an unstoppable feedback loop in which lingering patches coalesced into snowfields, which coalesced into a cap of ice thick enough to flow outward under its own weight, making it a glacial ice cap. Over thousands of years, the local ice cap enlarged into a great dome of ice flowing outward in all directions along rays distorted by the regional slope, not unlike a starfish with bent legs. This was the Labrador ice dome. A similar progression on the western side of Hudson Bay formed the Keewatin ice dome. By eighty thousand years ago, these two domes had coalesced with smaller domes to the north to become a unified mass.[8]

The Labrador and Keewatin domes were the "Adam and Eve" for every kettle in the glaciated fringe. Though physically distinct, they behaved as a couple, merging near southern Hudson Bay and diverting flow southward. New England, the Catskills, and western New York were fed by ice from the Labrador Dome, which crept and slid across the St. Lawrence River. From Minnesota to the west, everything was fed by the larger Keewatin Dome.[9] The area in between resulted from flows involving interactions of both domes, with the Labrador being dominant. Because both domes were nourished from the same North Atlantic source of moisture, a strong buildup of ice in Labrador would partially starve the Keewatin Dome of snowfall. Conversely, when the eastern dome thinned, Atlantic moisture could reach farther west, enhancing the flow from Keewatin. This symbiosis helps explain why the edge of the southern ice sheet was always readjusting its flow patterns and why the details of the chronology vary so much from east to west. Geologists long ago gave up trying to interpret Laurentide behavior as a straightforward result of global climate change. Too many codependent factors were involved.

This first attempt by the Laurentide to create kettles in the United States miscarried. About eighty thousand years ago, the Laurentide reached the St. Lawrence River, where it stalled and reversed course. Melting continued. Hudson Bay was reexposed. Whales returned. The western dome disappeared. Caribou returned to graze. About sixty thousand years ago, however, climatic conditions shifted back to those favorable for growth. This time, the push was stronger. Laurentide ice oozed across the Canadian border to invade the United States. It kept growing until it reached its maximum southern limit about twenty thousand years ago, before melting back to inaugurate the present interglacial epoch.

These comings and goings were preordained by planetary astronomy. Earth's elliptical orbit around the sun is slightly off center. Every hundred thousand years it stretches out to become slightly more eccentric, then returns to a more circular path, changing the duration of summer relative to winter. This slight change was the main pacemaker for global ice ages. Within the last hundred thousand years, the Laurentide's slow advance, recession, strong readvance, and disappearance were timed by two other orbital components. The tilt of Earth's axis relative to the plane of its revolution around the sun varies a few degrees with a rhythm slightly longer than forty thousand years. This increases and decreases the contrast between winter and summer. The wobble of Earth's spin axis mentioned earlier has a frequency near twenty thousand years and was the third orbital component involved. The wobble changes the timing of the equinoxes, either reinforcing or partially canceling out the other influences. In every case, seemingly trivial changes in solar radiation were greatly amplified by other components of the climate system. The stars seem to have been aligned during the Laurentide culmination.

The outermost margin of the Laurentide is marked in many places by a terminal ridge formed where the forward-moving blob stalled before reversing direction. Louis Agassiz, a Swiss scientist who established the glacial theory with *Études sur les glaciers*, called such ridges moraines.[10] The word is derived from the Latin *morena*, for "bank of stones," and some moraines are just that, places where boulders were dumped after being released like bales of hay at the end of a conveyor belt. More often, they are low ridges of sand and gravel.

Where terminal moraines are absent, the ice limit is marked by a dramatic change in soil type. Beyond the margin in the humid eastern United States are brownish pink, clay-rich, stone-poor soils produced by prolonged weathering. Beyond the margin to the west are thick, dark brown grassland soils richer in loam. North of the ice margin, the topsoil is much thinner, and the subsoil is usually either sandy or composed of a dense, stony soil called till, the residue of glacial crushing of rock deposited directly by the ice.[11] Early natural historians called it "boulder clay" owing to the puzzling blend of lumpy boulders and more finely ground-up material mixed with rock fragments. When pasted to the land by smearing beneath the ice, it is called lodgment till. Early American farmers called it "hardpan" because, having been compacted by the weight of the ice and given a stubborn grain by horizontal shearing, digging through it was nearly impossible.

The most subtle evidence for a former Laurentide margin is a single exotic boulder carried down from the north and parked above or amid more mundane materials. Such far-traveled stones are called erratics, after the Latin *errare*, meaning "to wander," because it's hard to know where they came from.[12] Many erratics are locally famous due to their colossal size and precariously balanced positions. They are particularly well exposed along the shorelines of kettle lakes because giant boulders had a tendency to tumble into the growing melt depressions, and later wave action exposed them.

In New England, the ice sheet began to withdraw from its outer limit as early as twenty-five thousand years ago.[13] Within five thousand years, the net result of all the thrusting, buckling, and dumping taking place at its edge was an archipelago of sandy islands between Nantucket and New York City, each a segment of moraine buried by a wash of meltwater sand and gravel. After several feeble readvances, the glacier profile flattened, then died mostly downward rather than backward. The southern part of the region was free of ice by eighteen thousand years ago, the coast of Maine before thirteen. Highlands in the northern Appalachians held a small residual ice cap for a few thousand years after that.

The basins of Lakes Michigan, Huron, Ontario, and Erie began to deglaciate about the same time New England did. But about eighteen thousand years ago, the ice margin readvanced strongly, refilling the basins and expanding beyond them. For the next five thousand years, streams of ice came and went within each basin like commuter trains at a busy station, sometimes moving at speeds of up to a mile per year. These oscillations occurred at a millennial scale governed by internal changes associated with meltwater production beneath Hudson Bay. By twelve to thirteen thousand years ago, the edge of the Laurentide Ice Sheet had retreated back into Canada, except for a final readvance across the northern part of Lake Superior. So cold was the deglacial climate of the High Plains that it likely held the last block of buried glacier ice in the United States, which probably didn't melt until about nine thousand years ago in northern North Dakota. The Keewatin and Labrador domes remained confluent above Hudson Bay until about seven thousand years ago. With further shrinkage, the sea returned between them, effectively ending the last glacial epoch. A trivial part of the Laurentide remains today as an icy fossil beneath the Barnes Ice Cap on Baffin Island.

ICE SHEETS ARE A WEIRD PHENOMENON—each is a hybrid cross between Earth's water cycle and its rock cycle. Though precipitated from the atmosphere as snow and rain, the Laurentide can also be considered an enormous blob of soft rock that spread outward under its own weight. Every grain of its ice met the criterion for a mineral, and every aggregation of mineral grains meets the criterion for a rock, although the melting temperature of ice sheets is about two thousand degrees Fahrenheit below that of normal rocks and their strength far below that of sandstone and shale.[14]

A pile of ice less than about a hundred feet thick is a brittle solid, which is why the upper and frontal surface of every glacier is crevassed. When it is thicker than that, however, the pressure causes the ice to behave plastically, and the pile begins to bulge outward like dough or putty. This internal movement occurs by a slow, internal shearing mechanism called creep, which is similar to what happens when a deck of cards is pushed forward from the top. This is the only mechanism available when the glacier is frozen to its bed. But when liquid water is present at the contact between land and ice, two additional possibilities arise. On a hard, rigid bed such as granite, forward motion takes place by sliding. On a bed dominated by glacially pulverized residues, motion occurs by forward smearing, similar to a knife spreading chunky peanut butter on a cracker.[15] Both of these mechanisms are enhanced by hydraulic pressure, which helps lift the load, and therefore reduces friction with the bed.

The forward motion of the Laurentide Ice Sheet was powered by the continuous gravitational flow of excess mass in the interior deposited as snow and a deficit near its margin removed by melting. When the balance was positive, meaning that there was a net gain in the total amount of ice, the edge of the blob advanced southward. When it was negative, the edge retreated northward. But the ice was always moving forward like some kind of slow-motion landslide that never quit, at least until it thinned below the threshold required for motion.

Because the Laurentide was a mass of ice rock, its bed can be considered a geological fault zone.[16] With normal rock, the frictional bond is usually strong, and movement occurs in dramatic lurches, accompanied by powerful earthquakes. With ice rock, however, the bond is weak, and fault motion takes place in a series of tiny jerks, each accompanied by a low-magnitude icequake. When glaciers move forward by smearing over their beds, however, fault motion is continuous. Under these conditions, the noises picked up on seismometers are the snap, crackle, and pop of stones being broken into bits.

Laurentide ice rock didn't have the strength to crush stones or erode the Canadian Shield by itself. Instead sharp slabs of rock fragments embedded in the forward-moving ice gouged, scratched, scraped, and flaked each other and the surface. Grit and stone debris incorporated into the basal ice made it not only darker but stronger, allowing even greater force to be applied. The whole crushing system was efficient because the basal fault zone was constantly being rinsed clean by meltwater moving under pressure.

When exposed, glacier ice exhibits many colors and textures. If from deep inside the glacier, it can be crystal blue. Refrozen slush is milky white or gray. Tar black or chocolate brown ice comes from what geologists call the dispersed zone, where grit and ice are blended like pigment in paint. Zebra stripes result from bands of debris-rich ice separated from pure ice.

The boulders that tumbled out of the glacier more than twelve thousand years ago—granite, slate, metal ore, gneiss, basalt—were lucky survivors of the crushing process. Earth's continental crust is warm and soft at great depth. Near the surface, it cools, expands, becomes more brittle, and is squeezed by the vise of horizontal tectonic forces. The inevitable result is a three-dimensional set of brittle fractures called joints, which intersect to produce discrete blocks of rock with geometrically regular but angular shapes. Joints were opened, widened, and loosened by weathering as the ice sheet approached, especially when infiltrating water froze in the cracks and expanded. Breakup continued after the ice sheet covered the rock due to local strains, increased hydraulic pressure, and elastic flexure of the crust. Blocks were yanked forward when ice melted on the upstream sides of obstacles and refroze downstream.[17] Well back from the margin, blocks moved upward and downward through chance collisions and where ice was being added or lost from the bottom. Within a mile or two of the ice margin, however, they moved mostly upward by compressive thrusting and buckling. Upon reaching the glacier surface, they became concentrated as the ice disappeared, and as rain, snowmelt, and glacier melt washed the sand and grit away.

Near the bottom of the shearing ice mass, the largest and strongest blocks of rock rolled in slow motion, becoming partially rounded lumps. Medium-sized slabs got caught up in the heavy traffic just above the bed. They scratched and abraded the bedrock, jostling against one another and becoming either milled into more rounded forms or broken into more jagged ones. Unless exceptionally hard or lucky, fist-sized or smaller fragments were readily crushed to a residue. Gritty rocks like mudstone and limestone were pulverized back into the silt-sized particles from which they were made.[18] Fine-grained crystalline and cryptocrystalline rocks like basalt or flint broke down along a continuum of particle sizes. However, the granite and gneiss so typical of the Canadian Shield and the northern Appalachians were aggregates of strong crystals at the millimeter scale ($\frac{1}{32}$ inch), separated by weaker bonds. Quartz was the strongest common crystal, both mechanically and chemically. After crushing, the result was a residue dominated by quartz-rich sand, mixed with larger and smaller fragments.

Rinsing the residue was the final step in the production of kettle lake sand. The Laurentide melted downward at a rate of several feet per year over a broad marginal zone up to several hundred miles wide. To this surface meltwater was added seasonal snowmelt, seasonal rainfall, and the water produced by frictional sliding and geothermal heat. On the ice sheet interior, surface melt percolated into snow and was usually refrozen. Nearer the margin it flowed within surface channels until it found a fracture or crevasse and plunged downward to the bed. There the pressure rose, pushing water toward the edge as some combination of thin films, networks of shallow channels carved into the bed, and tunnels melted up into the ice. Squirting

*Cleanly rinsed glacial sand near Lake Nebagamon, Wisconsin.*

from the margin, especially in late spring, was a steady gush of water, silt, sand, pebbles, cobbles, and boulders.

The boulders and cobbles required high stream velocities to move. Once beyond the glacial margin, the flow weakened, and they were quickly left behind. Silt and clay didn't settle out until the flow became tranquil, usually within a lake. Grains of medium and coarse sand, however, were easily picked up by water because the particles were big enough to catch the flow and small enough to be rolled and bounced along the bed by the drag of moving water. The result was that a formerly heterogeneous mixture of particles was sorted into distinct size classes at progressively greater distances from the ice edge.

The zone dominated by deposition of sand would play the most important role in creating kettle lakes and ponds across most of the glaciated fringe. Within it, lenses of brown gravel were present where meltwater channels cut downward, and seams of gray silt were present where grit settled from turbid slack water. But glacially crushed, well-rinsed, pebbly sand was by far and away the dominant material.[19] It's typically off-white in color due to the abundance of silica in its source rock. But certain minerals give rise to pastel tints. Pure pinks come from the red-bed hematite ores of the Superior Upland. Orange tints come from the crushed feldspars of Canadian Shield granite. Dusky brown is found in the iron-rich sands of New England's volcanic provinces, olive brown above the Cretaceous shale of the Dakotas, creamy tan in the limestone country of upstate New York, greenish in the lightly steamed slates south of Providence and Boston, and plain shades of gray practically everywhere else.

# The Birth of Kettle Lakes and Ponds

THE MOST WELL-KNOWN CONTINENTAL divide within the United States is the mountainous north-south line separating west-flowing Pacific streams from those flowing east over the plains. The more important continental divide in terms of American history runs roughly east-west, approximating the border between Canada and the eastern United States. It separates north-flowing drainages of the Hudson Bay–St. Lawrence River system from south-flowing drainages of the Missouri-Mississippi-Ohio system. This divide was crossed back and forth during the seventeenth century by the French explorers who laid claim to Louisiana. During the eighteenth century it was the battleground of the French and Indian War. And during the nineteenth century it separated English fur traders of the Hudson Bay Company from those operating in the Old Northwest Territory of the fledgling United States.

This east-west continental divide also played the decisive role in creating the band of kettle lakes across the northern United States. The Canadian Shield, the most ancient and stable part of North America, is also a broad bedrock basin centered on Hudson Bay. It can be thought of as an immense, shallow bowl that rises southward to a rim approximating a broad arc from Maine to Minnesota. The five Great Lakes and the dozen or so lesser "great" lakes to the west— Winnipeg, Athabasca, Great Slave, Great Bear, and beyond to the McKenzie River—are low spots on the rim of the bowl, through which tongues of ice leaked southward. These are not kettles, because they were carved from rock beneath actively moving ice. Rock-carved lakes are also common in the northeastern United States. For example, the large lakes of the Maine wilderness—Moosehead, Chesuncook, Rangeley—are ice-gashed bedrock basins over thirty miles long. So are the "almost great" lakes: Winnipesaukee, Squam, and Sunapee in New Hampshire, Lake Champlain on the Vermont–New York border, and many others. The Finger Lakes of west-central New York are also rock carved. They are deep gouges bordered by flat-lying rocks of limestone, sandstone, and shale, which is why waterfalls are so common there.

The topographic bowl containing the Laurentide was even more pronounced during its culmination. Directly beneath the shield, the earth's crust is a cold, stiff layer more than twenty miles thick atop the mantle, a much thicker layer that is warm enough to deform like exceptionally stiff taffy. As the ice sheet grew in size, its dead weight was transmitted downward through the rigid shell of the crust into the doughlike mantle beneath it. The result was slow outward flow beneath the bowl, creating a giant depression more than a half-mile deep. Beyond the rim of the bowl was a broader, upward bulge, which made the depression seem even deeper.

The Laurentide reached its highest point somewhere near the deepest part of the bowl.[1] From there, ice and meltwater flowed southward, and slightly uphill, fanning out between New England and the Dakotas. Having reached the edge of the bowl, however, Laurentide ice was able to move forward as a thinner mass and spread out. Basal meltwater followed the same pattern, moving outward from the center of the bowl, where the pressure was highest, draining away and dropping its load just beyond the edge. Sediment being carried forward within basal layers of the ice was also concentrated just beyond the edge of the bowl.[2]

The basins of the Great Lakes acted as escape valves for the continent-sized pile of ice to the north. Buoyant lobes of ice within them flowed faster and farther than the thinner ice on adjacent highlands. During ice retreat, the resupply of ice to the interlake highlands became more difficult. Additionally, lobes were guided toward each other until they collided. All of these processes—a change in bed slope, thinner ice on highlands, spreading meltwater streams, and lobe collisions—facilitated stagnation of ice between the Great Lakes and large valleys and the subsequent burial of the stagnant ice blocks by sand.

In New England and in the Dakotas, the ice sheet had a second rim farther to the south. From New York City to beyond Nantucket was a north-facing escarpment up to three hundred feet high marking the edge of the now-submerged coastal plain. In the Dakotas was the Missouri Couteau, the low, north-facing edge of a remnant plateau. When the Laurentide encroached on these escarpments, it did the same thing it had done in the heartland, which was to compress, slow down, and thicken in order to climb up and over them. This set the stage for the pattern of moraine formation, stagnation, and copious meltwater production that created the many kettles of the sandy archipelago of southernmost New England and the greatest concentrations of prairie potholes to the west.[3]

During deglaciation, the enormous mass that had previously been pushing the mantle outward disappeared. As the mantle returned, it lifted the crust upward. Hudson Bay is still rising, because the rate of recovery is slow and because the outward push lasted nearly six times longer than its inward return. The bay will eventually be lifted high enough to drain completely, provided the ice doesn't come back. Meanwhile, everything to the south of a hinge line along the northern shores of Lake Huron and Lake Superior is sinking. The net effect was to flatten the bowl and return the craton to its preglacial equilibrium.

KETTLE LAKES AND PONDS continue to form today near the margin of the Malaspina Glacier on the northern Gulf of Alaska coast. Nourished by some of the highest mountain snowfalls on Earth, ice fields drain down to the coastal plain, where the ice puddles up and spreads out into an enormous lobe similar in size and behavior to the ones that created Cape Cod Bay and Long Island Sound. Present are moraines, ice stagnation topography, small glacial lakes, and pitted outwash plains bearing a striking resemblance to those deposited across the glaciated fringe after the Laurentide culmination. The Malaspina provides a modern analogue for what took place in the glaciated fringe between about twenty thousand and twelve thousand years ago.[4]

Malaspina meltdown was reported during the eighteenth century by Danish explorer Vitus Bering (sailing for Russia), British captains James Cook and George Vancouver, and Spanish commander Alessandro Malaspina, for whom the glacier is named.[5] In 1794 Vancouver described rock debris collapsing downward above melting ice in the terminal zone and a geographic transition from pure ice to "loose unconnected stones of different magnitude," then to tundra, then to forest. In 1837 Captain Sir Edward Belcher figured out where the surface debris was coming from: "veined and variegated streaks of debris within the ice." In 1848 Russian hydrographer Captain Michael Tebenkof was eyewitness to modern Malaspina kettle formation at a time when his scientific contemporaries were debating the origin of similar features in Ohio based on geological evidence.[6] Shortly after the turn of the twentieth century, the National Geographic Society sponsored pioneering geological expeditions to the Malaspina, which fully documented the meltdown sequence leading to kettled terrain. More recent studies based on aerial photography, satellite imagery, and field observations confirm the relevance of using the Malaspina as a model for kettle lake formation across the glaciated fringe.

I have seen Malaspina kettles in the process of being formed, but only from the air. My first opportunity to observe the meltdown sequence on the ground was at Lituya Bay on the Gulf of Alaska coast, originally named Port des Français in 1786 by the explorer Jean-François de Galaup, comte de La Pérouse. There the entire bay is rimmed by lateral moraines on which enormous conifers collapse like matchsticks into growing kettles. I've also watched them grow and stabilize in the northern foothills of the Alaska Range, where alpine glaciers descend eighteen thousand feet below the summit of Mount McKinley in Denali National Park.[7] In the terminal zones of these glaciers, jagged cliffs of crevassed ice lie beneath thick mantles of debris. On moraines that are about three thousand years old, ponds of various sizes and shapes are enlarging amid hills of brushy tundra dominated by dwarf birch, willow, and alder. I've seen kettles in the sedge tundra of the Anaktuvuk and Sagavanirktok valleys of the northern Brooks Range that began forming more than sixty thousand years ago, before the meltdown sequence was put on hold by permafrost. Only recently, and in response to recent global warming, has the meltdown sequence continued toward its inevitable conclusion.

The burial of stagnant Laurentide ice happened automatically when the glacier readvanced over blocks of ice trapped in depressions, bulldozing and smearing sediment. In these circumstances kettles are bordered by bouldery till. Burial of blocks by a sheet of looser, jagged debris happened near prominent terminal and recessional moraines because compression at the edge forced bands of debris to shear up over masses of stationary ice. During downward melting, the debris spread out, leaving a chaotic terrain with many small, stony kettles. If the shearing was localized, the result was a cigar-shaped kettle where an ice-cored moraine once stood. The vast majority of Laurentide kettles, however, were created when slabs of ice became stranded during retreat, like stragglers falling behind a crowd, and were then buried or surrounded by the steady wash of sediments being sluiced down from the north.[8]

The normal process of kettle birthing began when south-creeping ice thinned to the point where forward motion ceased, leaving a sheet of stationary ice several hundred feet thick or less. Where the sheet was well drained and the supply of sediment low, it thinned into residual blocks, leaving little trace of the meltdown process. More often, irregular blocks of ice halfway through the meltdown process were buried—perhaps in a decade or a century—by a broad fan of sand and gravel being flushed away from icy tunnels to the north. The result was a cluster of isolated lakes with jagged shorelines below a fairly smooth sand terrace similar to those encountered by the Pilgrims at Truro. These are called pitted outwash plains.[9] If the ice stagnated beneath the water level of a glacial lake, the block was usually buried more deeply, giving rise to rounded kettles separated by ridges at different elevations.

Melt craters developed in every circumstance, regardless of the source of the sediment. Where the ice was thin, the craters quickly bottomed out and migrated backward, producing ice-walled ponds and leaving low plateaus of lake sediment. A much more chaotic meltdown scenario happened when a thick layer of stony debris or water-laid sand lay over a thick slab of stagnant ice. Stony debris slid into melt craters. Muddy sediments oozed into them like wet concrete. Sandy sediments washed downward. Boulders toppled into the growing holes. Then, in a randomized reversal of fortune, what had been holes became hills, because the sediment-filled depressions were better insulated. This topographic inversion repeated itself at a declining pace until the ice was completely gone. During this process, more and more of the clay and silt washed away, increasing the looseness and permeability of what remained. The final result was an irregular, well-drained surface of holes, hollows, hills, and ridges called ice stagnation terrain. The holes and hollows are kettles. Notable hills are called kames, the tallest and most distinct of which occur where streams on the glacier surface plunged into a crevasse, filling the hole with sand and gravel. Eskers are also composed of sand and gravel, but are ridges, the remains of either open-air channels flowing between ice blocks or subglacial tunnels that became backfilled before the ice disappeared.

The main feature of ice stagnation terrain is its chaotic topography. In 1878 geologist T. C. Chamberlain used the term kettle moraine to describe a broad band of this topography. Here, he describes one in southeastern Wisconsin:

> The superficial aspect of the formation is that of an irregular, intricate
> series of drift ridges and hills of rapidly, but often very gracefully,

undulating contour, consisting of rounded domes, conical peaks, winding and occasionally geniculated ridges, short, sharp spurs, mounds, knolls and hummocks, promiscuously arranged, accompanied by corresponding depressions . . . One of the peculiarities of the range is the large number of small lakes, without inlet or outlet, that dot its course.[10]

His literary chaos matches that of the moraine. Henry Rowe Schoolcraft, another nineteenth-century polymath, paddled through the grander, flatter version of ice stagnation topography while searching for the headwaters of the Mississippi River. He wrote that the streams travel in practically every direction and described a large kettle named Leech Lake as "one of the most irregular shaped bodies of water that can be conceived of . . . a combination of curves, in the shape of points, peninsulas, and bays, of which nothing short of a map can convey an accurate idea." Swedish author Vilhelm Moberg wrote of the almost fractal pattern of a kettle lake complex in Minnesota: "a conglomeration of islets, peninsulas, points, inlets, bays,

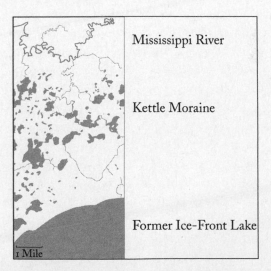

*Geographic patterns near Aitkin, Minnesota.*

necks, headlands, isthmuses." So crazed is the map pattern of lakes in this area that they are mythically alleged to be the footprints of Babe the Blue Ox, Paul Bunyan's stalwart animal companion.

The shape of Lake Chargoggagoggmanchauggogoggchaubunagungamaugg, a kettle in Webster, Massachusetts, is only slightly less confusing than its name.[11] Thoreau was stunned by the topographic randomness of kettle moraines on Cape Cod, comparing them to a "chopped sea." Henry Beston visited the same region a half-century later, calling it a "belt of wild, rolling, and treeless sand moorland." Pioneering archaeologist William A. Ritchie described the kettle moraine on the south edge of New York's Finger Lakes near Corning as having "numerous tiny lakes and ponds . . . an undulatory configuration to the sand and gravel knolls." To me, this topography resembles chaotically twisted muffin tins or three-dimensional labyrinths dominated by curves rather than lines. Geographers use the word "deranged."

Some of the randomness in the distribution of lakes disappears at larger scales. In some cases, blocks of ice were preferentially detached in preglacial valleys, yielding a pattern resembling lumpy beads on a string. More commonly, irregularly shaped lakes and large clusters line up along broadly curved arcs. These mark places where the edge of an ice lobe stagnated after a strong advance, focusing the detachment of blocks and the sand to bury them. Wisconsin has one such arc hopscotching southeast from Lake Mendota to Lakes Monona, Waubesa, Kegonsa, and Koshkonong. The Detroit Lakes area of western Minnesota has a broad fan of slightly aligned lakes extending northwest and southeast of Otter Tail Lake. Michigan has a less distinctive sweep from Round Lake to Crooked Lake, with at least fifty small lakes in between. Iowa, Illinois, Indiana, Ohio, and New York all claim a "Chain o' Lakes" of some magnitude caused by the alignment of kettles in a moraine.

The soil of ice stagnation topography is almost universally dry and sandy, supporting patchy forests of pine and oak and with an undergrowth of drought-tolerant grasses and shrubs. It is often of limited agricultural value, used for firewood and hay. On the prairie, kettles are usually flanked by bluestem grasses, wildflowers, sage, and even cactus, and the topography is considered an impediment to "big-sky" agriculture. When ice stagnation topography develops in the finer-grained soils near glacial lakes, it retains enough moisture to be productive pastures.

Finally, ice stagnation terrain has little in the way of surface drainage. Thoreau called Walden a "deep green well." Naturalist John Muir was more specific, saying it was "fed by currents which ooze through beds of drift." This is indeed true for most kettles, which are fed by cold groundwater springs. Schoolcraft remarked that the Mississippi headwaters country of north-central Minnesota "abounds in pure springs, and is so impervious in its lower strata, that it has probably retained to the present day, more water in the character of lakes, large and small, than any other part of the world." In this regard, the true source of the Mississippi River is not Lake Itasca but the vast groundwater realm that feeds it.

Over most of the glaciated fringe, but especially in New England and the Great Lakes states, the underground flow takes place through sand crushed from quartz-rich rocks like granite, gneiss, schist, and quartzite. This makes the water chemistry "soft" and slightly acidic relative to the more "hard-water," alkaline conditions produced by limestone, other marine sedimentary rocks, and basalt. Hence, kettles have a tendency toward greater chemical purity than that of small, comparably sized bedrock lakes with inlets. Isolated kettle ponds find their chemical and botanical equivalent in bogs, whereas bedrock lakes find their equivalents in marshes and fens.

The final stage of ice stagnation and kettle formation began when all traces of surface ice had disappeared and the remaining ice was at or below ground level. Geothermal heat was being conducted steadily upward from the endless supply of the earth's interior. Surface heat was reaching slowly downward during summer, moving through the insulating layers of sediment. Surface heating was more efficient for blocks of ice beneath water because the liquid was less well insulated and because it absorbed and held more heat than adjacent soil. The pace of ice-block melting decelerated with time as it became more removed from surface heat, sometimes dragging on for thousands of years. Exactly when the last ice disappeared would have been difficult to tell from surface observations.

Physical evidence for the successive stages of the meltdown process is still visible under some circumstances, such as in glaciated mountain valleys where braided gravel-bed rivers migrate laterally into moraines, in coastal settings where the sea erodes bluffs backward, and in gravel quarries where excavating machines dig into the banks. In such eroded and excavated places, small kettles can be seen in cross section. At the top, a layer of peat usually overlies water-laid silt and clay that, in turn, overlie a loose jumble of sediment created by meltdown subsidence and flowage into melt craters. At the base is usually a layer of till or scoured rock over which the ice was sliding before it became motion-less.[12] Toward the margins are beds of sand and gravel broken by gravity faults created where masses of sediment slumped downward as support of the ice block was removed. The bottoms of broad kettle hollows are often more bouldery than elsewhere because the lingering slab of ice prevented subsequent burial by sand and gravel.

WHEN THE LAURENTIDE ICE SHEET crept southward over the edge of the Canadian Shield's great topographic bowl, it produced kettles in five large geographic sectors set off from one another by topography: Northeastern, Great Lakes, Superior, Dakota, and Montana. Each sector had its own glaciological personality and distinctive pattern of kettle formation.[13]

The Northeastern sector covers all of New England, New York City, nearly all of upstate New York to the northern tip of the Allegheny Mountains in the western part of the state, the top of New Jersey, and the northeast corner of Pennsylvania. East of the Hudson River is mostly rough topography developed on the hard crystalline rocks—granite, gneiss, schist, slate—of the Appalachian basement.

There the ice margin was a wavy series of lobes, one for every large bay or valley: South Channel, Cape Cod Bay, Buzzards Bay, Narragansett Bay, the Connecticut River Valley, and the Hudson River Valley. Sediment deposition and moraine formation were most intense in the areas between lobes, which took the shape of an upside-down V opening to the south. The best example is from Martha's Vineyard, Massachusetts, where a pitted outwash plain opens southward to famous Katama Beach. Fifteen miles to the north, the main part of Cape Cod is a clone of this pattern. Its apex opens southward to the kettle-pitted Mashpee plains above Falmouth.[14]

On the rougher topography of the New England mainland were two contrasting modes of kettle formation. Where the main ice sheet thinned above south-draining bedrock watersheds, enormous masses of ice were cut off from resupply by the drainage divide to the north. As these masses melted downward, the blocks that lingered longest were those located in the downstream centers of each valley, where the ice was originally thickest. There sediment from the adjacent hills and slopes flushed downward onto the narrowing and thinning residual blocks. To this was added sediment washing down each valley from larger unmelted slabs to the north. The net effect was to bury narrow ribbons of ice beneath broadly braided rivers. The final result was elongate valley kettles flanked by flat gravel benches called kame terraces, the edges of which are often heavily kettled. One such place is Mansfield Hollow, in Mansfield, Connecticut, which is a long, irregular meltdown depression flanked by sand plain on both sides.[15]

In north-draining watersheds like the Sudbury River in Concord, Massachusetts, a different pattern developed. There high-level glacial lakes were impounded between a receding ice dam to the north and a notched bedrock spillway to the south. Purgatory Chasm in Sutton, Massachusetts, and Devil's Hopyard in East Haddam, Connecticut, are spectacular examples of these meltwater "flumes" cut across resistant rock; both are easily accessed as state parks. Each has potholes drilled into solid rock by stationary eddies caught in a rush of meltwater so strong that boulders were swirled in whirlpools until they were ground down to nothing. On the northern side of each spillway-controlled lake, meltwater gushed out from beneath the ice to build massive deltas over whatever blocks had become stranded in bedrock hollows. This is how Walden Pond was created. Thoreau understood that the high terraces on opposite sides of Walden were once part of the same high-level delta, which he called a "diluvial" surface. He also cited the Algonquin creation myth for the pond, in which a hill rose "as high into the heavens as the pond now sinks deep into the earth . . . [and the] . . . stones rolled down its sides to become the present shore."[16] This is exactly what happened. The debris-covered hill of ice became a hole, the stones rolling downward during collapse.

Kettles farther west in New England follow a similar south-to-north variation. Block Island and the famous Charlestown area of Rhode Island are moraines similar in age to those on Martha's Vineyard and Cape Cod, respectively, and which also formed between two lobes. Thousands of kettles were produced on the moraines and the sand plains draining from them. Those on the Charlestown Moraine are the stoniest in my experience, owing to the ice-thrust origin of the surface debris. The Ronkonkoma and Harbor Hill moraines form the backbone of Long Island. They correspond in age to those farther east and were created by more sprawling lobes in the Connecticut and Hudson river valleys. The highest point on the island is located where moraines from both advances and from both lobes converge. This is no coincidence.

West of Staten Island, across the top of New Jersey, and along the northeastern flank of the Alleghenies in Pennsylvania, the ice merely nosed its way into countless mountain valleys, never finding enough room to create large lobes or enough sediment to build big moraines. Kettles in eastern Pennsylvania and upstate New York mimic the inland New England pattern of north- vs. south-draining watersheds, especially in the Leatherstocking terrain west of the Hudson and south of the Adirondacks. Something different happened on the flat-rock platform of central New York to the west. There, after pulling back from the Alleghenies, the Ontario Lobe developed a broadly curved margin, advanced, deepened the Finger Lakes from their previous conditions, and deposited prominent recessional moraines at both ends of the lakes. The result is dozens of kettle lakes decorating both ends of the larger bedrock gashes.


NEARLY A THOUSAND MILES west of the Alleghenies lies the Superior Upland, an ancient, tough-as-nails, bumpy plateau reaching slightly less than two thousand feet in elevation in the Huron Mountains of upstate Michigan near Marquette. This upland partitioned south-flowing ice in the basins of Lakes Erie, Huron, and Michigan from west-flowing ice in the basin of Lake Superior. To the east lies the Great Lakes sector, the most glaciologically temperamental part of the glaciated fringe. It covers westernmost New York, the adjacent corner of Pennsylvania, most of Ohio and Indiana, the northeastern third of Illinois, the eastern half of Wisconsin, and nearly all of Michigan. The smoother topography, generally softer rocks of the platform, and more buoyant conditions led to the formation of six major lobes, and gave the ice sheet there a much lower surface slope than in adjacent sectors, causing fast glacier flow and producing countless recessional moraines.[17]

As with New England, the Great Lakes sector also exhibits a strong north-south contrast in its kettles. The south is dominated by the great moraine loops of central Illinois, central Indiana, and Ohio. These moraines are usually several miles across, rising gently above either flat lake plains or the undulating till plain. Most of the material isn't sand at all, but loamy silt and stony clay pasted to the land. This material was so easily eroded by subglacial meltwater that a shallow network of shifting channels usually developed, rather than a single ice-roofed tunnel, minimizing the potential for ice-block burial and kettle formation.[18] And the kettles that did form were more likely to have been drained by gully erosion, filled by gully deposits, or smothered by a mantle of windblown fine sand and dust.

This dust, called loess, is responsible for the fertile soils of the corn-soybean belt from Iowa to southeastern Ohio. Though deposited by wind, the dust source was glacial in origin. During winter, meltwater tunnels to the north would squeeze nearly shut because there was less water flowing through them. Each spring the combination of rain and snowmelt would raise the water pressure north of the constricted tunnels. At some critical point, the tunnels would quickly enlarge, and the water would gush out in a great annual flood of sand, silt, and gravel. When this flood receded, the mud deposited on high gravel bars and abandoned channels dried out. Prevailing westerly winds then blew this grit eastward as coarse dust, burying kettles beneath a blanket of loess more than a hundred feet thick near the Mississippi River, its most important source.

During the Laurentide culmination, ice streaming southward through the Lake Huron basin kept right on going in the same direction, advancing across the basin of Lake Erie to the Appalachian foothills in southern Ohio and Indiana. Dozens of sublobes, one in every north-draining valley, created hundreds of kettles just inside the outermost limit. As deglaciation continued, however, the flow was channeled west-southwest into a fast-moving lobe flowing parallel to Lake Erie. As this tongue of ice expanded and contracted, its southern edge rose and fell against land, creating a discontinuous band of kettles extending all the way from westernmost New York through western Pennsylvania, central Ohio, and northern Indiana. Collisions between this fast-flowing lobe and the Lake Michigan Lobe were responsible for the highest concentration of kettles in Indiana and Illinois. When the Erie Lobe retracted, the Lake Michigan Lobe was able to spread southeastward into the Hoosier State. When the Erie Lobe was more extensive, it pushed the Lake Michigan Lobe westward. The sharpest kink in this collision gave rise to the lake district north of Chicago in Cook County.

Within the northern part of the Great Lakes sector, the main kettle lake action took place on the highlands between or surrounded by lobes. The Lake Michigan Lobe lay parallel to one in Green Bay. In the gutter between them accumulated a linear kettle moraine nearly two hundred miles long. It's responsible for most of the lakes in eastern Wisconsin. The highlands between Lakes Erie, Huron, and Michigan are also rimmed by kettle moraines. Lower Michigan and southwestern Ontario were surrounded on all sides.

So strong was the divide between the south-flowing ice in the Great Lakes sector and west-flowing ice in the Lake Superior sector that an "ice shadow" more than two hundred miles long developed. Within it lay the "driftless" area of Wisconsin, Minnesota, and Iowa. During the maximum advances of Pleistocene ice, this was a lozenge-shaped hole of nonglacial terrain surrounded on all sides by ice. During the Laurentide culmination, it was a long peninsula indenting the ice margin from the south. Here there are no kettles, moraines, or natural lakes. Instead, the topography and soils more closely resemble those of unglaciated central Kansas, with sandstone buttes and "chimney" rocks.

The Superior sector lies west of the Huron Mountains divide, covering the western half of Michigan's Upper Peninsula, most of northern Wisconsin, and the eastern half of Minnesota north of the Twin Cities. Geographically, it's the smallest sector because ice flow was shunted at nearly a right angle to the main flow heading south. Ice had to crawl up and out of the deepest part of the deepest lake, in excess of two thousand feet at the time. Despite its limited area, this sector produced the greatest number of kettle lakes, especially in northern Wisconsin and Minnesota, the "land of ten thousand lakes," which has the densest concentration in the United States. This was a collision zone between multiple lobes coming from multiple directions during multiple ice advances.[19] Ice from Labrador being channeled generally southeast pushed back and forth against ice from Keewatin, traveling either southwest, northeast, or southeast. The Lake Itasca area was the main collision zone, into which meltwater poured from all sides, piling up hundreds of feet of sand and gravel.[20] The Mississippi River literally springs from the chaos of this multiple confluence.

Northwest of Lake Superior is a remote, largely unoccupied hinterland. There, somewhere between Upper Red Lake and International

Falls and protruding through one of the largest bogs in the United States, lies the divide between the Superior and Dakota sectors. West of the divide, Keewatin ice was channeled southward in two great lobes: the Des Moines Lobe and the James River Lobe. Moving over mostly clay-rich shale, these lobes left richer soils than those left by the Superior Lobe, which traveled over the Canadian Shield. In jest, author Garrison Keillor remarked that if the Des Moines Lobe had arrived first, "Lake Wobegon's history would be much brighter than it is."[21]

In the southern part of the Dakota sector, clay-rich soils helped restrict kettle formation. This is low, rolling, loamy country with small moraines arranged in a corrugated process and with few kettles. To the north, moraines are of only modest size, but they seem like mountain ridges on this otherwise "horizontal world."[22] Being sandy and stony, they were generally avoided by farmers except for the "borrowing" of gravel. Some became skiing destinations in places where the relentless wind didn't blow the snow away. Even farther north in the Dakota sector, the ice likely advanced over permafrost, which helps explain the widespread stagnation and the generation of thousands of prairie potholes. The shallowest are simply low spots on the former undulating till plain. The most unusual are "pop-ups," formed when the glacier froze onto its bed before lifting up a scab of sediment. The deepest and most perennial potholes are small kettles, made all the more beautiful by their isolation.

The western limit of the Dakota sector is marked by the Missouri Couteau, an elevated prairie plateau rising in South Dakota near the Nebraska border. From there it follows the shape of a boomerang, north then west, before fading away in northwestern North Dakota. At the center of this curved highland lies Bismarck. John Steinbeck wrote:

On the Bismarck side it is eastern landscape, eastern grass, with the look and smell of eastern America. Across the Missouri on the Mandan side, it is pure west, with brown grass and water scorings and

*A prairie kettle near Sharon, North Dakota.*

small outcrops. The two sides of the river might well be a thousand miles apart.[23]

They are indeed far apart, geologically speaking. The glaciated landscape to the east has newer, fresher, richer, and moister soils, speckled with kettles and potholes. To the west, the more ancient alkaline soils of the prairie have been developing in place for tens of millions of years, and natural lakes are virtually unknown. On the Couteau itself is a concentrated band of prairie kettles and other potholes where south-flowing ice compressed, concentrating debris on the surface. Ice and debris then melted down into chaos.[24] Because the bedrock there consists of soft shale instead of crystalline rock, the kettles are often muddier than elsewhere in the fringe. In one place, a Mud Lake lies adjacent to a Big Muddy Lake.

To the west lies the Montana sector, which follows an east-west ice limit between Williston, North Dakota, and Great Falls, Montana, before arching north to the Canadian border. There the southerly push of the ice was weaker, giving rise to a glacial limit marked not so much by moraines as by the present path of the Missouri River, whose course was established as an ice-marginal stream. Near the Canadian border, the clay-rich prairie landscape was streamlined into long ridges and grooves. A few prairie pothole kettles are present where the ice stagnation topography developed. One glacial phenomenon unique to the Montana sector was the creation of many small ice-free islands on locally elevated plateaus. They are the flat-topped prairie equivalents of nunataks, jagged peaks that rise dramatically above mountain glacier systems.

# 3.

# Extinction, Early Humans, and Stabilization

ACCORDING TO A NATIVE AMERICAN MYTH, while traveling in a white canoe over water of exceptional clarity, the Ojibwe magician Chebiabos saw "heaps of beings who had perished before, and whose bones lay strewed on the bottom of the lake."[1] Archaeologists have indeed found bones from gigantic, shaggy ice-age creatures strewn on the bottoms of North American kettle lakes and ponds near their shores. Roaming the landscape were mammoth, mastodon, stag moose, beaver, long-horned bison, lion, cheetah, ground sloth, tapir, peccary, camel, horse, dire wolf, condor, and most frightening of all, the giant short-faced bear.[2] Typically the bones lie either within or upon a basal layer of mineral mud capped by softer, peaty sediments of organic origin.[3] This boundary and the fossils upon it bear witness to the extraordinary environmental transition between glacial and interglacial worlds.

Poet Walt Whitman lamented the loss of the ice-age animals, writing: "In vain the mastodon retreats beneath its own powder'd bones."[4] Animal skeletons were indeed pulverized into oblivion within the sandy, well-drained, acidic soils above kettle lakes and ponds. But below the water surface, fossils of all kinds have been chilled, pickled, tanned, salted, sealed, and soaked into preservation. Kettles are extraordinary archives for past life, the most important fossil repositories in eastern North America.

Nowhere is the fossilization better than in so-called trash layers, found from Cape Cod to North Dakota.[5] Hardly trash, each is an intact layer of peaty soil sandwiched between glacial materials at the bottom and lake sediments above. With kettles, this layered sequence required that the ground surface be stabilized and vegetated before the stagnant ice melted to create a lake basin. During the meltdown any bone left on the surface would have been chilled by oxygen-poor cold water, entombed by the enveloping sediment, and deeply buried.

Purely physical processes dominated kettle lakes and ponds during their first several millennia, given the colder temperatures, greater seasonality, increased exposure to stronger winds, limited ground cover, deeper frost heaving, thicker lake ice, and fissures associated with the final stages of meltdown collapse. But the force of life was already there. Bacteria were living between sand grains. Diatoms—beautiful versions of single-celled algae with glassy silica "skeletons"—were drifting in ultraclear water. Lichens, a cooperative blend of fungi and algae, were encrusting shoreline boulders. Stands of sedge and horsetail rush crept out over wave-washed cobble pavements. Patches of heather, dwarf willow, poplar, and alder were colonizing protected hollows. A few kettles had inherited fish species from earlier glacial lakes, perhaps char, grayling, or lake trout. Mastodons, barren-ground caribou, and nomadic human hunters were beginning to pass by, perhaps stopping for a drink.

This physical stage of aquatic history left its signature in lakes as a distinctive layer of light-colored, fine-grained mineral sediment lining the bottom of practically every water-filled kettle in the glaciated fringe.[6] The luster and texture of this "silt liner" is that of pottery not yet fired. In fact, when it was later exposed by erosion, Eastern Woodland tribes excavated the silt liner to make their primitive ceramic pots. European colonists dug it for bricks, stoneware, mortar, and caulking. In small kettles often aptly named "clay pits," the shape of the liner resembles the bell of a brass musical instrument like the French horn.[7] In larger kettles, the liner takes the shape of a broadly flaring saucer. The liner is better developed in kettles than in isolated bedrock lakes of equivalent size because the weaker bank materials of the former provided a stronger sediment source for a longer time.

Sand grains brought to a lake by wind or water will be deposited close to the shore because they sink so rapidly.[8] Silt- and clay-sized particles, however, disperse throughout the lake as a turbid suspension that clarifies only under calm conditions. Calmest of all is the water beneath a thick layer of winter ice. Protected from wind, and coinciding with a time when streams are either dry or frozen, the volume of water within each lake basin becomes nearly motionless for months, allowing even the finest clay to settle. Mud that settles in shallow water is easily resuspended by wave action and animal activity the following year. But particles settling in deeper water below the zone of surface turbulence go down for good. There is simply no mechanism to get them back up.

The silt liner is the perfect marker for the ice-age transition. Unlike the collapsed glacial material at greater depth, it's perfectly respectable lake sediment. The darker organic lake sediment above it—called detritus by lake scientists and muck by nearly everyone else—dates from the younger, more biologically productive phase of the present interglacial period. The shift from physical to organic sediment was due largely to the successful colonization of ice stagnation terrain by plants. Roots went down to anchor soils with a living mesh. Stems, dead leaves, and fibrous residues armored the terrain and facilitated the infiltration of rain rather than runoff. As shrub tundra replaced carpet tundra, the increased aerodynamic roughness sapped wind energy. When trees returned, the flutter of leaves and the flexing and swaying of branches sapped even more. Bluff top forests deflected winds to higher levels, stilling waves on the surface. The combined effect of forest growth on the land was to calm the water surface, reduce storm flows, and greatly limit the supply of silt and sand entering open water.

THE DISPERSAL OF PLANTS over the upland landscape and the succession from tundra to forest were also associated with another, more widely known transition, the extinction of the shaggy ice-age megafauna. Thirty-five genera of terrestrial North American mammals disappeared in what was arguably the most dramatic extinction since the end of the dinosaurs. Woolly mammoths—the most charismatic species—were common during Laurentide recession, especially in the southern Great Lakes states.[9] These oversized grazers preferred the cold, dry, and generally treeless conditions of the continental interior, especially Wisconsin's driftless area in the ice shadow south of the Superior Upland. Mastodon, which were more adapted to browsing and wetter conditions, were more numerous and more widely distributed toward the east. The predators and scavengers dependent on these and other herbivores vanished along with their prey. Barren-ground caribou, also known as reindeer, were once widespread from Maine to Iowa. They survived by following the edge of the ice northward. Today they are the most common large herbivores in the subarctic.

All three transitions—animal extinction, plant colonization, and the mineral-to-organic shift in lake sediment—came together in kettle marshes and bogs. On the surface was a dense mat of herbaceous and woody roots woven together. In the subsurface was a saturated fibrous muck, sometimes with the consistency of jelly. These would have been dangerous places for heavy creatures coming to drink or to graze, natural traps for ice-age life, equivalent to the soft soils of Kentucky's salty Big Bone Lick, the steep-sided sinkholes of Florida,

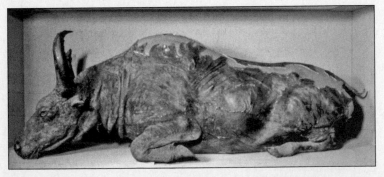

*An extinct ice-age bison from permafrost near Fairbanks, Alaska.*

and the sticky Rancho La Brea tar pits of Los Angeles. Kettles also often have steep bluffs that merge in narrow bays, not unlike the box canyons of the west used by Native hunters to drive game. The combination of soft soils and steep bluffs would have made some kettles good ambush sites for early human predators going after large, heavy animals. Though perfectly plausible, such a connection has not yet been convincingly made.

Even the most skeptical archaeologists, however, are convinced that humans butchered a mastodon on the edge of Pleasant Lake, a small marshy kettle in south-central Michigan.[10] Eleven to twelve thousand years ago, a carcass was cut up and the prime parts hauled away. The evidence for butchery consists of the pattern of cut marks on the bones and the separation of carcass parts. The evidence for a lakeshore setting comes from beds of shoreline silt and sand overlain by peat full of spruce cones that likely floated into place. The mastodon may have been killed there, or perhaps was simply scavenged. More interesting, but less convincing to skeptics, is the nearby Heisler site, where, mastodon expert Daniel C. Fisher claimed, humans submerged a mastodon carcass in the acidic, cold water to help preserve it, using stakes and stones stuffed into the beast's intestines to weigh it down.

The most well-known group of these Paleo-Indian hunters were those of the Clovis tradition. Restricted to a four-century-long slice of time between 13,200 and 12,800 years ago, theirs was the most geographically widespread and most culturally uniform archaeological tool kit in the New World.[11] Their diagnostic artifact was a beautifully symmetrical, lance-shaped spear point, skillfully fluted on one end for hafting, and found in an area bounded by New England, Guatemala, California, and Alaska. These weapons were clearly lethal because they are associated with the bones of mammoths. Other Clovis artifacts made from stone, bone, and ivory are mostly simple scraping and cutting tools. Figurative art, symbolic representations, and permanent house sites are unknown, suggesting a people constantly on the move, and with one main thing in mind: meat.

The ecologist Tim Flannery wrote with great flourish that "the Clovis frontier was a meat frontier, and that when the last mammoth steak was finished at the last Pleistocene megafaunal barbecue, so too were Clovis and the North America megafauna."[12] The majority of paleontologists, however, believe that the dramatic environmental changes taking place at that time—spasmodic glacial advances, the freakish warm-cold oscillation known as the Younger Dryas, a transition from dusty to moist soils, and a destabilized ecosystem unlike any before or after—were largely responsible, with human hunting playing either a subsidiary role or no role at all. Mammoths and mastodons lingered for another millennia or two after most other ice-age creatures disappeared, as did the short-faced bear, a horrific obligate predator, standing twice as tall as a grizzly and with twice the weight. Since the bear was capable of running faster than twenty-five miles per hour and overlapped in time and space with Paleo-Indian hunters, it most likely had a few human meals.[13]

Many years ago, I was so curious to taste ice-age meat that I actually tried some raw. I had been excavating for fossils in a bank of permanently frozen pond sediment near Chicken, Alaska, when I encountered the frozen knuckle of an extinct steppe bison (*Bison priscus*) with some jerkylike tissue still attached. After rinsing it off in a nearby stream, I took a bite, chewed, and swallowed. It wasn't very good.

Buried bones and archaeological remains are much more common than the limited number of sites suggest. This is because almost all such discoveries are accidental, found only when the overlying layer of muck—often more than ten feet thick—is being excavated or explored for some other purpose, usually construction. The Pleasant Lake site, for example, was discovered in the bucket of a dragline. The silt liner is also seldom seen because it, too, lies deeply buried. But it's assuredly there, visible to every geologist who has seen kettles cut in half by erosion and to every pollen scientist who has pushed one of their sampling tubes deeply enough.

· · ·

KETTLE LAKES ACROSS THE GLACIATED fringe stabilized long before European exploration. This stability, however, was a very long time in coming, especially near large bodies of water. In southern New England many kettles that are ponds today were formerly dry hollows, even after meltdown was complete. It took many millennia for them to fill with groundwater. Other kettles took the opposite course, holding water early in their history before losing it to the tides of time. Understanding the coming and going of water in these basins gives lakeshore residents some perspective on what is often taken for granted: the permanency of their body of water.

Think of groundwater as a vast underground lake existing not as a single large void, but as billions of tiny ones connected between sediment grains or rock fractures. The boundary between saturated conditions at depth and unsaturated conditions higher up is called the water table. Seldom perfectly flat, it rises with infusions of snowmelt and rain, and falls under conditions of warm, sunny skies when evaporation and plant transpiration coincide with steady underground drainage. Above the table, water moves by trickling downward, like rain dripping through a tree canopy.[14] On the surface of the table, the flow is sideways and gently downhill, always toward the nearest stream, lake, or sea. Beneath the table, water moves under pressure, as does water in house pipes being fed by gravity from a tank. Though the infusions of water come rapidly from rainstorms and snowmelts, their drainage takes a long time because the water is forced to flow sideways long distances through tortuous pathways. If it never rained and if shorelines were fixed, the gentle slope of the water table would eventually flatten to become as horizontal as a lake surface on a calm day.

Over the span of decades, water table elevation fluctuates around an average condition. However, on a scale of centuries and longer, the average height of the water table can rise or fall quite a bit. For example, the postglacial rise of sea level against Long Island raised the water table more than three hundred feet. Climate also controls the level of the water table. If the average weather gets wetter or drier, the regional water table can rise or fall tens of feet, increasing or decreasing the number of lakes, respectively. Rivers can build their channels up or cut them down, and since they are the natural drains for groundwater, these fluctuations also change the level of the water table. The earth's crust can tilt up in one place and down in another, forcing lakes in or out of existence. This was especially important for kettles adjacent to the Great Lakes and Lake Champlain in Vermont.

Kettles in the sandy archipelago of southern New England formed exactly as did their counterparts in the heartland. But two surprises were in store. Over the millennia, sea level rose against the shore, flooding the continental shelf and raising the water table below dry land. What had been empty kettle holes filled with freshwater to become pools, then ponds. The second surprise came during the last few millennia, after the pace of sea level rise slowed. This gave Atlantic storm waves time to erode sandy moraines backward to produce prominent bluffs fronting broad beaches. As the line of erosion migrated backward, there were moments when the edge of the stormy sea met the edge of a sandy kettle, breaking the barrier between them. Freshwater rushed out. The sea rushed in. What had been a freshwater pond became a biologically rich saltwater estuary separated from the sea by a bar of cobble or boulder gravel. Salt Pond, in Eastham, Massachusetts, once a freshwater kettle, is now a cul de sac estuary at the head of a short marine inlet.[15] Kettles breached by the sea even earlier were sliced in half, making the scalloped bays that are commonly present behind barrier beaches, especially on Nantucket Island. Some were breached so long ago that they've been erased, except for their peaty and silty remains submerged beneath the shoaling sands of the sea.

The story of Cape Cod kettles reminds us that the final geological step creating kettle lakes and ponds anywhere within the fringe is their filling with groundwater. Kettle holes and hollows are bone-dry, regardless of size. They hold water only during the wettest days of spring or after heavy rainstorms, and occur only above the water table, typically on the highest, driest parts of sandy moraines.

Kettle pools hold water for only part of the year. The small ones are seasonal "vernal" pools that cup water above the silt liner. They fill to capacity with spring rain and snowmelt. During the summer, the water slowly disappears, usually from evaporation and plant transpiration. Often no bigger than a backyard swimming pool, each is a tiny oasis of wetland life within otherwise dry soil.[16] Large kettle pools usually form in the zone where the groundwater table oscillates seasonally up and down, and in situations where the silt liner is shaped like a saucer or platter. Normally, they become swamps.

Ponds are perennial because they are almost always fed by seepages from the groundwater reserve.[17] The smallest ponds look like tiny pools or dug wells. Ponds may be as large as ten acres, according to the U.S. Environmental Protection Agency's arbitrary cutoff—any larger and they are called lakes. Colonial settlers on the Atlantic seaboard followed the English practice of using the term "lake" for large, rugged, rock-bound bodies of water. Almost everything else was called a pond, regardless of whether it was freshwater or salt, big or small, natural or man-made. The construction of water-powered mills—used to provide mechanical power for cutting logs, grinding grain, and other activities—added to the confusion over terms because the body of water raised above every milldam was called a pond, regardless of size or appearance. Therefore, some lakes became ponds, even as they were enlarged. Colonial Massachusetts issued a ruling requiring that privately managed farm ponds measuring less than ten acres be differentiated from larger, publicly managed bodies of water called Great Ponds, most of which were naturally formed. Over the years, Great Ponds were reclassified as lakes, though seldom with a change in popular name, which explains why so many New England lakes have the same name: Great Pond. This also explains why seven of the nine largest legal lakes in Massachusetts are officially called Ponds with a capital "P." Walden Pond is the deepest lake in the state and has four times the surface area needed to qualify as such.[18]

Attorneys, legislators, and land-use regulators like the arbitrary distinction between lakes and ponds because it's so straightforward to use. Lake scientists do not. They prefer to distinguish ponds from lakes based on criteria that actually matter to the ecosystem. For them, there is little difference between a water body that is 9.99 acres and one that is 1.01 acres. One definition of a pond that lake scientists like is: a place where the water is shallow enough to allow wind mixing from top to bottom, thereby preventing the formation of stable thermal layers. Fifteen feet is a common depth at which this happens. Alternatively, ponds are defined as bodies of water either shallow enough for rooted aquatic vegetation to cover the entire bottom or too tranquil for beaches or wave-washed rocky zones to develop.

BONES AND STONE AGE TOOLS are the most intriguing remains found within kettle lakes and ponds; however, they are neither the most common nor the most important. We learn much more from the leaves, wood, and seeds of trees and shrubs, insects, and aquatic invertebrates that fall directly into the water, float outward, and sink. Even more important are the microscopic grains of pollen carried in by the wind from various distances. The abundance and ubiquity of pollen more than compensate for its tiny size.[19]

Pollen is the sperm of the vegetable kingdom. Practically all of the higher plants reproduce sexually by producing pollen on the male sexual organs (stamens) and receiving it on the female organs (pistils). Many specialized plants rely on insects and birds for distributing their pollen. But most of the trees, grasses, sedges, and flowers cast their fate to the wind, producing pollen in astonishing quantities in order to ensure that at least a few grains fall on a waiting pistil, often hundreds of miles away. To enhance wind transport, pollen grains have become extremely light in weight and have developed various shapes to catch the wind. To enhance survival, each is covered with a tough polymer coating called an exine.

As tough as they are, most of the pollen grains that fall on the surface of the earth are either destroyed or jumbled together. Pollen that falls on soil is readily decomposed by bacteria because oxygen is available. The bottoms of swamps and marshes lack oxygen and therefore preserve pollen, but most are poor pollen archives because plant roots mix the mud continuously and because their peat deposits decompose when the water level drops. In shallow water, the pollen record is complicated by the nesting of fish, the burrowing of invertebrates, and the stomping around of large herbivores like moose. Large lakes usually have problems associated with the resuspension of pollen by waves and the receipt of pollen from inflowing streams.

In short, almost every place in the world is a bad place to obtain a long-term, high-fidelity record of the local pollen rain. Kettle lakes provide a rare exception, one in which the chronology is readily obtained by radiocarbon dating.[20] Pollen cores from kettle lake muck are the most important fossil archives for humid midlatitude regions, equivalent to pack rat middens of the desert. They are for paleoecology and climatology what cave sediments are for archaeology, and what deep sea cores are for glacial history. Boggy kettles have the added advantage of reduced water-surface tension, which helps pollen sink quickly, thereby maintaining a more accurate record.

Pollen science is mostly done in the laboratory. Long hours are spent digesting the resistant grains from the muck, counting them under the microscope, and converting these counts into statistically robust measures of former vegetation. Sampling kettle lakes for pollen is far more exciting. It's done with strong metal tubes pushed down into stiff sediment. These are equipped with detachable rods for pushing, a movable piston that retracts during the push to prevent shunting the sediment aside, and spring-loaded flanges to keep sediment from falling out when the tube is raised back up through the water. Large coring devices are usually set up on winter ice. Equipped with heavy weights, vibrating motors, tall tripods, winches, and cables, some resemble miniature oil derricks. When a lake that doesn't freeze deeply is sampled, a floating platform is either anchored or tethered in place. Core segment after core segment, scientists and technicians push the sampler down, pull it back up, extrude a segment for laboratory analysis, then go back down for another refill. Unless stopped by friction, they eventually reach the much stiffer and lighter-colored silt liner. This tells them that they've reached the bottom of the biological archive, back into the physical world of the earliest lakes. If they went deeper, they would encounter either the collapsed sediments deposited during meltdown or the layer of till or bedrock over which the ice slid before coming to rest.

Most lake muck has the consistency of a fibrous gel. Pollen scientists call it *gyttja*, a Swedish word pronounced "yitt'-yuh." Neither liquid nor solid but something in between, it ranges in texture from stiff pudding to soggy shredded wheat. Much of it consists of dead phytoplankton. Practically every larger speck has been torn, macerated, chewed, shredded, ingested, and passed through the gastrointestinal tract of some creature, often more than once. Present within the *gyttja* are unseen variations in the concentration of silt and clay. Sometimes there are thin layers of mineral sediment caused by exceptional floods, shoreline erosion, or landslides. In bogs acidic enough to limit microbial life at the bottom, the muck is so soft that a weighted fishing hook will sink right through it. To sample such "false bottoms," pollen scientists resort to a coring device refrigerated with dry ice that is inserted to the bottom before being rotated open, allowing the soft gel to freeze onto the tube.

The tree pollen within kettles of the glaciated fringe is dominated by spruce, birch, pine, and oak. Nontree pollen is dominated by sedges and forbs, a term used by botanists to describe grasses, most wildflowers, and nonaquatic herbaceous plants. Each of these major plant types can be unambiguously related to a modern environmental setting through a sampling of the modern pollen that falls today. For example, sedge pollen falls in high concentrations today only where moist, cold, windswept tundra is widespread. Forbs are diagnostic of the drier, more continental climate of western prairies. Spruce pollen is plentiful in the continental boreal (meaning northerly) forests of Canada. Cool and moist conditions of the north woods are signaled by birch. Oak dominates the eastern half of the United States north of the Gulf Coast, indicating mid-latitude warmth and moderate humidity. Pine is important in the southeastern United States inland from the Gulf of Mexico and in the northern and western Great Lakes states.[21]

Kettle pollen records start in the Northeastern sector at about fourteen thousand years ago, the earliest transition between the physical and biological stages. At that time, and because of an extra-cold Atlantic chill, southern New England was dominated by sedge tundra, particularly on Cape Cod, Martha's Vineyard, and Nantucket. A barren-ground tundra plant known as *Hudsonia* was common, indicating moist, cold, windy conditions similar to those of Hudson Bay. Within the next few thousand years, vegetation converted to a mixture of sedge, spruce, and birch in which pine became increasingly dominant.

Meanwhile, in the more continental climates of the Great Lakes and Superior sectors, a sedge-spruce parkland was characteristic. It converted to a mixture of spruce and pine with spruce being most common to the north. There was also an east-west gradient, with forbs being codominant to the west and sedge to the east. Later, all sectors east of the Dakotas were dominated by pine, and west of there by forbs. Vegetation changes during the next several millennia were complicated by an abrupt climate reversal toward colder conditions.[22]

By about ten thousand years ago, the glaciated fringe was getting more summer sunshine than today. But the lingering ice sheet to the north remained a significant climatic force. Its blinding reflectance more than made up for the extra dose of radiation that arrived. It poured chilly meltwater into the Atlantic, which impacted temperatures throughout the hemisphere. It still held enough water to keep sea levels low, keeping maritime moisture at bay and allowing expanded coastal plains to reflect away more light than water would. Its topographic dome, though lower, remained capable of creating a cold summer anticyclone from which drained katabatic winds. The combination of gusty winds, cold water, and strong seasonality aerated the water of kettle lakes, helping to keep their bottoms clean and sandy.

By nine thousand years ago, however, the ice sheet had melted back too far and had become too flat to exert much influence on the glaciated fringe. Streamflow and groundwater had become less alkaline, now that the most easily dissolved substances had been lost.[23] Plants with slower migration rates had generally caught up with those that had raced ahead in the first few millennia. Specks of bronze mica, pastel feldspar, and black iron-rich minerals began weathering into clay and oxide minerals that were better able to capture and release moisture and nutrients.[24] The growth and dieback of plant roots and the burrowing activity of many creatures began to generate a darker, more loamy topsoil over what had once been a lag of stone and sand.

A soil ecosystem of microbes, fungi, and countless invertebrates within the loam became a metabolic factory capable of extracting ions from mineral surfaces, binding them with organic matter, and cycling elements from winter to summer. These processes captured and held organic nutrients, particularly nitrogen and phosphorus. As the soils began to reach their storage capacity, water draining over and through them began to carry excess nutrients. Waiting next in sequence, at least in terms of elevation, were wetland soils, which captured and held their share of nutrients within peat and dense living biomass. Only when terrestrial and wetland sites had reached their fill of nutrients trickling down from above was the nutrient flux substantial enough to foster aquatic plant growth beyond the minimal condition. The productive, stable biological phase of kettle lake history had begun.

A LAKE WITHOUT PLANTS would be biologically sterile, something like a desert. In a deep lake with a narrow shore, the dominant plants are microscopic one-celled algae. These phytoplankton are invisible when the level of nutrient is low because they are widely dispersed in the water and because the green color of chlorophyll is masked by the slightly greenish tinge of lake water. In shallow lakes or those with broad, gentle margins, photosynthesis is often dominated by visible aquatic plants, most of which are rooted to the bottom. Though called weeds, many of these plants are fascinating wildflowers, with special adaptations for underwater germination and seed dispersal.[25] Other so-called weeds are special types of multicellular algae similar to the kelp of the sea. Still others are special types of ferns and mosses adapted to underwater life.

Herbivorous zooplankton are miniscule animals that eat the microscopic plants. In turn, they are eaten by slightly larger carnivorous zooplankton no larger than grains of sand. In turn, these are eaten by minnow-sized carnivores, which are eaten by medium-sized fish, and so forth, right on up the line to pickerel, eagles, and otters. With each step up this food chain, much of the stored chemical energy is lost, though that which remains becomes more concentrated. The base of the food chain is more complex nearer the shore, where a host of nibblers, biters, shredders, and scrapers from practically every animal phylum dine on large plants and each other. They also become food for the "higher-ups."

The term "food chain" works pretty well as a metaphor when the system is simple, meaning that one type of creature dines almost exclusively on one below it, and is, in turn, dined on by one above. But under normal circumstances, there is no linear pecking order of who eats whom. "Food web" becomes a much better metaphor because it connotes multiple connections in space and time, and emphasizes the importance of scavenging. Indeed, all lake life, whether visible or invisible, is part of a strong, flexible, and integrated web of energy captured from the sun. Life and death, though important to the individual organism, are irrelevant with respect to aquatic ecology, except as a handoff of chemical power.

Lakes were alive during their early physical phase, when the silt liner was accumulating. But the current of biological energy was weak. Low productivity and high oxygen levels meant that few aquatic fossils were preserved. Eventually the nutrient levels across the glaciated fringe rose to the threshold where more plant tissue was being produced than consumed. At this point, kettles became archives not only for pollen, but also for a variety of aquatic fossils. Most important are diatoms, photosynthetic algae that are very sensitive to water clarity, temperature, and chemistry and that preserve nearly as well as pollen. Taxonomic oddities like single-celled amoebas with armor plates (freshwater foraminifera), tiny crustaceans the size of seeds (ostracods), shrimplike crustaceans (water fleas), the head capsules of worms responsible for macerating detritus into muck (chironomids), and the hard covers of insect wings (elytra) all provide clues to these ancient aquatic ecosystems. They allow scientists to reconstruct the water depth, nutrient status, and ecosystem behavior of any sampled pond for any time in the past. Geochemists have also gotten quite clever about reconstructing lake histories, using indicators like stable isotopes, plant pigments, and mineral precipitates. Geologists reconstruct changes in the physical system, especially changes to paleo-shorelines and now-buried streams. The sum total of these aquatic records complements the terrestrial record told by pollen from the same archive.

After the late-glacial transition, lakes and ponds in every sector found their own appropriate level of biological activity, buffered by their surrounding wetlands and the thickening layer of detritus on the lake bottom. Whether a lake ran biologically lean or rich depended on whether there was a dearth or abundance of nutrient, respectively. Colder lakes tended to run leaner than warmer ones. Those in watersheds dominated by granite and limestone ran with less nutrient than those draining marine shale and volcanic lavas. Those fed principally by groundwater, rather than inlet streams, were less rich. Large, deep lakes in small watersheds ran leaner than small, shallow lakes in large watersheds. No two lakes were biologically alike, which is what made the population of kettle lakes so important to the Stone Age foragers who had begun to visit them regularly. Each lake had something different or the same thing at different times.

For the remainder of postglacial time there would be no abrupt regional-scale transitions to upland soils, forest ecosystems, or lakes within the glaciated fringe. But gradual changes were taking place, especially the arrival of new tree species and shifts in the range limits of others. The northern part of the glaciated fringe became a thermal battleground between the colder, coniferous north and the warmer, deciduous south. Truly northern forests of spruce, birch, and jack pine competed against truly southern forests of oak, maple, elm, ash, and ironwood, respectively.[26] Simultaneously, the southern part of the fringe became a moisture battleground between drought to the west and humidity to the east. True prairies and oak savannahs rich in grasses and wildflower forbs fought against forests richer in hickory, chestnut, and walnut.

Between about nine thousand and six thousand years ago, the plant communities better able to deal with heat and drought gained the advantage, pushing prairies northeastward into Wisconsin and oaks into Canada. Then, about four thousand years ago, the climate began to cool. This returned the advantage to the northern forests, which drove the prairies and oaks back where they are today. In this ecological battleground, the intentional burning by Native Americans and their role in controlling the distribution of grazing animals complicated the boundaries between forest and grassland types.

The effect of postglacial climatic variations on lakes is told not so much by the color and mineral content of their organic sediments as by the physical geological record of changing shoreline positions.[27] East of the modern prairie border, lake levels were generally close to their modern positions until about nine thousand years ago. Then, they dropped dramatically in response to the combination of warmer, sunnier summers and greater evaporation. They bottomed out between about five and six thousand years ago before rising back up to intermediate levels about three thousand years ago. West of the prairie border, lakes followed a similar but more dramatic pattern of changes. They fell to lower levels than in the east and returned back to higher levels before three thousand years ago. These shoreline records of water balance compare favorably with the abundance of charcoal in lake sediment—charcoal indicates fire, which indicates drought—and with microfossils indicative of water depth.

Falling lake levels lowered the water table, which allowed the aeration and decomposition of shoreline wetlands. Nutrients locked up in the peat were released, slightly raising the fertility of lakes at a time when the reduced lake volume was having the same effect. Conversely, rising lake levels of the last several thousand years had the effect of stabilizing (or in some cases slightly decreasing) the fertility of lakes by sequestering nutrients in wetlands, raising the volume of lake water, and increasing groundwater recharge. On those lakes where wave action by prevailing westerly winds was significant, the history of falling and rising lake levels is revealed by the erosion of broad wave-cut benches resembling miniature continental shelves. Resubmergence of the benches increased the habitat for rooted submerged aquatic plants, sometimes shifting the nutrient status toward higher fertility.

No two lake archives tell the same story. Some show oscillations in level, temperature, nutrient, and ecology at the decade and century scales. A forest fire indicated by a spike of charcoal might have been followed by a spike of fertility as rains leached the ash, and as water levels rose due to reduced transpiration by trees. Beavers may have raised the level of one lake but not the other, drowning wetlands and changing the nutrient flux. Alternatively, their dams may have failed, reversing the chemical balance. Cycles of herbivory and predation, microbial blights, insect infestations, and changing waterfowl migrations all influenced lakes. But on the whole, the lake-forest-wetland system stabilized and remained practically timeless, at least at the scale of millennia. It would remain that way until Europeans arrived.[28]

# 4.

# Natives and the Lake-Forest Ecosystem

THE HUMAN PREHISTORY of the glaciated fringe spans more than ten thousand years. On the human calendar, this is easily more than five hundred human generations. Native creation myths, oral traditions, lodge stories, and cultural memory provide anecdotal symbolic information for approximately the last thousand years, or roughly fifty generations. To understand what humans did before the earliest legends, we must rely on physical clues buried within the sediments and soils of archaeological sites.[1]

The deepest artifacts are the remains of Clovis big-game hunters and the bones of the creatures they hunted.[2] Their large and finely chipped spear points gave way to a variety of more diverse tool types, principally smaller side-notched bifacial points and ground-slate digging implements. The shallowest artifacts are European trade goods. Stone adzes gave way to steel axes. Crude pottery containers made of birch bark and woven reeds gave way to kettles made of iron. Gunpowder and bullets replaced bows and arrows. Trade beads replaced wampum, and broken whiskey jugs and glass bottles began to appear.

These archaeological findings correlate to the sedimentary record found within northern lakes and ponds. The disappearance of Clovis remains is associated with the transition between the mineral-rich, physically deposited sediments of the silt liner and the organic-rich biological muck from the present interglaciation. At higher levels, the disappearance of stone tools and their replacement by objects made of metal and glass coincides with abrupt ecological changes in many lakes caused by colonial activities: drowning from the construction of milldams, a dramatic increase in sedimentation caused by land clearing, and charcoal concentrations from accidental wildfires.

Between these Pan-American time lines, the story of kettle lakes and the story of the people who depended on them converge. Geologists refer to this entire postglacial interval as the Holocene epoch. Archaeologists divide it into four periods: the Paleo-Indian period, which ended about 8000 BCE (ten thousand years before present); the Archaic period, a preceramic, preagricultural interval still rooted firmly within the Stone Age, and extending until about 1000 or 500 BCE, depending on the definition;[3] the Woodland period, when native human populations became larger and their interactions more complex; and the period of sustained European contact, which began in the glaciated fringe about five hundred years ago with Jacques Cartier in 1534.

Most of the Holocene was dominated by quasi stability of the landscape.[4] Some lake ecosystems became slightly more productive with each passing millennium. And, as a whole, the population of lakes responded to the gradual change toward warmer, somewhat drier conditions during the first half of the Holocene, and to cooler and moister conditions during its second half. But at the century and millennial scales, the lake-forest ecosystem operated like a timeless, homeostatic machine.

Even before the beginning of the Archaic period, previously nomadic peoples began to settle down in regional groups and diversify. In the west, Clovis passed the baton to Folsom, a Paleo-Indian culture that became almost entirely dependent on bison. They were followed by the Plano, who retired the use of fluted points and developed the specialized hunting technique of driving game off cliffs into canyons and arroyos. Toward the east, Paleo-Indian groups began to regionalize as well. They adapted to hunting nonherding animals like moose, woodland caribou, and whitetail deer and foraging for food from rivers, lakes, the sea, and the forest. "At least seven distinctive Paleo-Indian fluted points styles are known from the Northeast alone."[5]

For the subsequent Archaic period, the archaeological record of the glaciated fringe—especially to the north—bifurcates along two main themes: continuous cultural change vs. a fairly stable pattern of finding food (subsistence). In site after site, and from one end of the fringe to the other, stone knives, scrapers, and projectile points show abrupt changes in material technology: the types of raw materials, the quarry sites from which they came, specific manufacturing techniques, and trade networks. New variations appeared so quickly and were so widespread that tool types are used as index fossils to subdivide the period. Dramatic technological change continued with the introduction of pottery and the cultivation of crops during the latest Archaic and early Woodland periods. Even more dramatic change is indicated by ceremonial objects and elaborate mortuary practices during late Woodland times. This phase culminated with the great effigy mounds of the Mississippian culture (AD 900–1300), which were built throughout the fringe and which overlapped in time with Viking explorations of the New World.[6]

When French, Dutch, Spanish, Portuguese, and English adventurers of the sixteenth century began to explore the glaciated fringe, they were stunned by the diversity of ceremonial customs and the polyglot of tribal languages that had spontaneously evolved.[7] The early-nineteenth-century geologist Henry Rowe Schoolcraft, considered America's first scholarly ethnographer, compared the resilience and distribution of tribes to the erratic boulders scattered by Laurentide ice during an earlier epoch, each far removed from its ancestral home. More recently, linguists have traced the use of shared sounds and grammatical structures backward in time to two great language groups: the Algonquin and the Macro Siouan, both of which were likely derived from a single common ancestor in the lower Mississippi and Ohio river valleys several thousand years ago.[8] As they dispersed north from an ancestral homeland, human bands moved into a terrain that had been geographically segmented by south-flowing lobes of ice. The result was an interweaving of distinct native groups across the fringe. The languages of the Malecites in Maine and the Ojibwe from Minnesota, for example, are descended from the Algonquin tongue; whereas the Winnebago of eastern Wisconsin and Iroquois groups of languages (Mohawk, Oneida, Onondaga, Cayuga, Seneca) of New York originated from the Sioux.

Contrary to some popular assessments of pre-Columbian America, native cultures in most of eastern North America deliberately managed terrestrial ecosystems with fire, streams and estuaries with weirs and canals, soils with agricultural plantings, and game stocks with selective culling.[9] This was especially true for the coasts and in large river valleys, where populations grew into ancient cities, the largest of which was Cahokia, located on the eastern bank of the Mississippi River in Illinois. There up to forty thousand residents, ruled by a priestly elite, built dozens of streets and hundreds of ceremonial mounds, one of which is larger than the Great Pyramid of Egypt. At less spectacular villages the food supply also depended increasingly on the intensification of agriculture, which remained supplemented by hunting and gathering.

In the kettled forests to the north, however, especially in settings removed from Atlantic shores and in climates with a growing season less than 150 days, the pattern of human subsistence seems to have remained nearly constant for most of Holocene time. This "timeless stability" is the second major theme in the archaeology of the glaciated fringe, especially to the north. The artifacts associated with food technology—spear points, scrapers, fishhooks, harpoons, net sinkers, food grinders, and so forth—varied in style and abundance, but their utilitarian function was constant. No new important food-procurement technologies were introduced, based on the stone and bone remains left behind. No important ones were discarded. For hundreds of generations, populations of *Homo sapiens* were simply one of many links in the stable food web, living within the ecosystem, rather than above it.[10]

ON THE SHORES OF LAMOKA LAKE—a beautiful kettle in upstate New York near the divide separating the St. Lawrence and Susquehanna rivers—the pioneering archaeologist William A. Ritchie found

> ancient camp sites on which the beveled adze is associated with celts, hammerstones, mullers, net sinkers, perforators, and narrow projectile

points. Pottery is always lacking and there is a paucity of bone implements . . . [These were] . . . temporary camps of a nomadic people . . .
Agriculture was unknown to the inhabitants, as was the fictile art, baskets and receptacles of wood, bark, and skins probably serving in lieu of
pottery. With the bow, javelin, spear, net, and hook and line, large meat
and fish supplies were obtained and smoked over great fires. Acorns
were certainly used as were doubtless nuts, berries, and various roots.[11]

His 1932 report came more than a century after Thomas Jefferson
inaugurated "pot-hunting" and "grave-robbing" archaeology with the
excavation of much younger Indian burial mounds in Virginia. Following Ritchie's investigation were hundreds of scholarly investigations that have refined and clarified the ancient lifeways of the natives
in eastern North America. Throughout it all, the Lamoka Lake site
has remained the standard reference for the Archaic period, to which
others were compared. Of the many sites located inland from the
coast and north of the glacial border, practically all are heavily dependent on lakes and wetlands, most of which occupy kettles.

*Lamoka Lake near Corning, New York.*

Every tool on Ritchie's inventory has to do with getting food or defending it. None have to do with ceremonial or symbolic aspects of life. The adze and celt were used to strip bark from trees and fell small ones, and as digging tools. Mullers were used to break up nuts and grind hard seeds. The ancient resident of Lamoka Lake who speared a hibernating bear for food was as much a part of the lake-forest ecosystem as the bear who ate the fish, who ate the minnow, who ate the mosquito larvae, who ate the water flea, who ate the diatom, whose nutrient came from bacteria dining on detritus. Every organism within this food chain, ranging from methane-producing bacteria to technologically sophisticated preagricultural humans, evolved within a single natural system, the complexity of which has been increasing for nearly four billion years.

Like all people during the Archaic period, Lamoka Lake Indians were skilled omnivores capable of manipulating their environment. They gathered and hunted food, returning to share it with the group, and competed for loosely defined territories. As with other animals, their populations were limited by the availability of food. Hence, they were as likely to starve as to thrive. Ecologically, Lamoka Lake Indians were neither "noble" nor "savage" because such terms are value judgments associated with human consciousness, rather than food procurement.[12]


A HUMAN ECOSYSTEM IS "a discrete human population that has a shared cultural inventory of technologic, social-organizational, and subsistence practices and is in interactive association with a specified environment."[13] Three basic developmental stages of the human ecosystem are generally acknowledged, starting with the phase in which humans were pioneering animals within an open, uncrowded environment. Social barriers with other groups, if present, were minimal; mobility was high; material technology was simple, highly portable, and oriented toward food gathering. Populations expanded quickly outward like the leading edge of a ripple on the surface of a pond. The Paleo-Indians fit this model: Their populations were low and mobile, their ecosystems were still adjusting to late-glacial migrations, and their technologies were experimental.

The second stage was the foraging version of human subsistence. Populations were typically of low density, dispersed over specific geographic areas, and generally on the move during seasonal rounds of food gathering. This was the pattern throughout the Archaic period within the glaciated fringe. Though the notion of home territories was clear, the concept of real estate with marked boundaries would have been alien to a group for which being on the move was part of the strategy for staying alive. Such peoples are called "hunter-gatherers" or "collectors."

The final stage was an agricultural society with settled villages, high populations, a material technology oriented toward food storage and ceremonial purposes, and fairly well-defined geographic boundaries associated with adjacent, often hostile groups. This was the pattern for the late Woodland period along the southern edge of the fringe and near the coast, especially during the warm spell that coincided with Norse explorations of the eleventh century. Tribes were dependent on

A Chippewa Family in the St. Croix Area, *by Sanford C. Sargent, 1885.*

the holy trinity of maize, beans, and squash, with the remainder of their diet based on wild game, fish, and gathered foods similar to those of the previous stage. The main villages were fortified with palisades, the fields were generally on stone-free floodplains and terraces, and the nearby countryside provided a host of other resources.[14]

The central and northern part of the glaciated fringe, however, was too cold for significant agriculture, especially during the last few centuries prior to European contact in the sixteenth and seventeenth centuries, which coincided with the Little Ice Age. This geographic swath coincides with the main belt of kettle lakes extending from eastern North Dakota through Minnesota, Wisconsin, northern Michigan, the shores of Lake Ontario, and eastward into the Maine woods. There the subsistence pattern of foraging prevailed not only during the Archaic, but during the Woodland and Contact periods as well, surrounded in all four compass directions by groups with contrasting material technologies and ways of life.[15]

To the north were the Shield Archaics, sparsely distributed nomadic bands that traversed the hard-rock, glacially scoured boreal landscape in search of caribou and fish. Their seasonal rounds covered hundreds of miles as they followed caribou migrations and fish runs, moving their camps on schedule with the animals. The technology was fairly simple: stone tools for hunting and fishing, fur clothing, and pedestrian transport, except for dogsleds in the winter. South of the fringe were the Central Riverine Archaics. These were more sedentary peoples of the unglaciated eastern deciduous forest, occupying villages and living off wild game, river mollusks, and a variety of vegetable foods. The Maritime Archaics lived east of Portland, Maine, on land with deeply indented rock shorelines and tidal rivers. They adapted to the tidal zone with a heavy dependence on protein from the sea. They hunted seals, scavenged whales, speared large fish in near-shore pools, netted smaller fish in bays, collected shellfish, and went after caribou in season. West of Minnesota were the Plains Archaics, who subsisted heavily on big game, especially buffalo, and lived in seasonal villages along the Missouri River, where they also caught fish and waterfowl.

Within the deciduous forests of the southern Great Lakes and along the Atlantic coast were the Mast Forest Archaics. This group, to which the Lamoka Lake site belongs, depended heavily on nut trees, particularly acorn, hickory, chestnut, and walnut. Their material technology was rich in tools for digging and processing these vegetable foods.

The Lake Forest Archaics occupied the cooler forests to the north, extending along a band parallel to the St. Lawrence Valley and the shores of the upper Great Lakes, and coinciding with the kettle lake heartland. Of all the Archaic groups, they are thought to have been the most flexible in their food procurement, with no known anchor protein from mammals—bison, caribou, marine mammal, or deer. Archaeologist Brian Fagan concludes that they subsisted "on a multitude of primary and secondary overlapping food resources. Deer and other mammals, medium-sized and small fish, shellfish, reptiles, waterfowl, ground-running and tree-roosting birds and all kinds of plants."[16] Obtaining this food required seasonal ranging over poorly defined territories.

The Lake Forest Archaics were heavily invested in fishing, especially after about six thousand years ago, when the rise in the number of sites coincides with intensive exploitation of aquatic resources.[17] Archaeologist Erhard Rostland demonstrated more than a half-century ago that tools associated with fishing—nets, hooks, and weirs—were ubiquitous, with harpoons and special fish-holding spears called leisters also being present at many sites. The presence of what appear to be decoys suggests that fish spearing was sometimes done through the ice within a darkened enclosure. "Of all implements, however, the fish net seems to have been the most important, suggesting a reliance on schooling and spawning fish."[18] Unlike native groups on the Northwest Coast or in Alaska who depended on a single type of fish (salmon), the Lake Forest Archaics exploited multiple species at different times and places. Archaeologists subsequently have confirmed Rostland's basic interpretations.

. . .

THE DESCRIPTION OF HUMAN CULTURES based on direct observation (ethnography) provides an important complement to New World archaeology for the period after European contact. When the Italian nobleman Giovanni da Verrazano sailed through a gap in the Laurentide terminal moraine between Long Island and Staten Island in 1524, he encountered a large society of Late Woodland natives in the former zone of the Mast Forest Archaics who were "clad with feathers of fowls of divers colors."[19] From there he sailed into Rhode Island's Narragansett Bay, where he spent several weeks with the Wampanoags, noting that they wore artistically embroidered deerskin clothing adorned with "divers stones of sundry colors" and "plates of wrought copper, which they esteem more than gold." In Verrazano's words, they "had no use for cloth—peltry suited them better—nor did they want iron or steel implements to replace their stone axes."

Ten years later, Jacques Cartier encountered Late Woodland natives in the former zone of the Lake Forest Archaics near Quebec. He remarked that they

> could well be called savages, for they are the poorest people that can be in the world; all their possessions, apart from the canoes and fishing nets, were not worth five sous. They wore nothing but a G-string and a few furs that they threw over their shoulders like scarfs . . . they eat both fish and meat almost raw, and their only huts are their canoes, reversed.[20]

Cartier makes the error of using material richness to gauge cultural richness. But his comment—along with hundreds of other early descriptions from sixteenth- and seventeenth-century historic accounts—does support the notion that the material wealth of natives at the time of contact was higher to the south and near the coast.[21] Those in warmer and more uniform climates where agriculture was possible had more material possessions, their populations were higher, and their territories better defined. Those living in the more boreal forests of the north, especially in the maze of kettle lakes in the northern heartland, were materially poorer, sparser, and more mobile.

Henry Rowe Schoolcraft's ethnography of the Ojibwe provides the best and most broadly applicable set of observations about native subsistence within the heartland during the Contact period. Born in 1793 in a small village west of Albany, New York, he was originally trained as a mining geologist. In 1822 he was appointed by the U.S. government as an agent for Indian affairs and posted to Sault Sainte Marie, Michigan, where Lakes Huron, Michigan, and Superior converge. This was effectively the center of the Ojibwe universe, the common ground from which they dispersed seasonally over the forested interior, especially to the south into the kettle lake belt. In 1823 Schoolcraft married the so-called northern Pocahontas, Jane Johnston, daughter of an Irish fur trader named John Johnston and Ozhaw-Guscoday-Wayquay (Woman of the Green Prairie), daughter of Ojibwe chieftain Waub Ojeeg.[22] Schoolcraft remained in Sault Sainte Marie until 1841. During his nineteen-year tenure, he learned the language and lore of the Ojibwe while living with his wife's extended family. He also traveled widely, often into the interior by birch-bark canoe.

Schoolcraft was certainly not the first to record observations about the northern "boulder tribes" way of life for posterity because resident natives were widely mentioned in the journals of voyageurs, explorers, adventurers, and Jesuit missionaries.[23] But he was the first to systematically collect a wide range of firsthand accounts from a single place within a scientific framework, to interpret the native mythology, and to publish his interpretations in scholarly monographs, books, and articles.

As noted in *Jesuit Relations* of 1640, the Ojibwe were first described under the name Baouichtigouin, "people of the Sault."[24] They had no written language other than pictures drawn on birch bark, hence European spellings of the tribal name were strictly phonetic attempts to translate sounds onto paper. After sorting through a variety of spellings—Achipoes, Ochipoy, Chepeways, Tschipeway, Otchipwe, Odjibwa, Odjibwe, Ojibway, Ojibwa, and

Ojibwe—the U.S. government settled on Chippewa for policy purposes.[25] Tribal members refer to themselves as the Anish-inaubag, the "spontaneous men" or "first people." Tribal lore suggests they migrated to the upper Great Lakes from the eastern Gulf of St. Lawrence during the Woodland period.

Schoolcraft worked during the first half of the nineteenth century, approximately four hundred years after Cartier's first descriptions of a closely related group. By that time, most of the original North American tribes had been severely impacted by Europeans, some to the point of extinction. Being spread out over interior lakes and woodlands, however, Ojibwe populations during the Woodland period had remained fairly low and mobile relative to those concentrated in fortified villages. This helped reduce the impact of contagious disease. Unlike eastern Algonquin groups such as the Pequots, or western groups of Siouan descent such as the Cheyenne, the Ojibwe were never subject to a government-sponsored campaign of genocide, perhaps because they were such valuable servants of the fur trade. Most important, the boreal lake-forest homeland of the Ojibwe was less desirable to the Europeans, whose principal interest was farming a place they could permanently settle. As a result, the tribe was more intact than any other within the fringe east of the Dakotas at a time when scholarly ethnography emerged. Their pattern of subsistence is most reflective of the ancient way of life that prevailed for most of postglacial time. Historian Bruce Catton described their home as a land where "nothing had happened since the glaciers melted."[26]

Schoolcraft characterized the Ojibwe as "wandering foresters," especially in Michigan, Wisconsin, and Minnesota. Though some groups did plant small plots of corn, they did so generally after European contact and to supplement their diets, rather than to sustain themselves, because depending on corn was too risky. Their housing technology consisted of wigwams or lodges generally similar to those of the east, but with more use of birch bark and reed mats of bulrush in lieu of elm or ash bark. Generally absent were heavy objects associated with the southern groups: for example, soapstone bowls, pestles, and mortars, which were too encumbering for a lifestyle of constant motion. Northern trappers never reported seeing Indians with heavy copper tools.[27]

Making a link between the foraging habits of the Ojibwe and their mental outlook, Schoolcraft supported the "creature of the forest" animal model for their human ecology, albeit with a heavy dose of cultural arrogance.

> The necessity of changing their camps often, to procure game or fish, the want of domesticated animals, the general dependence on wild rice, and the custom of journeying in canoes, have produced a general uniformity of life. And it is emphatically a life of want and vicissitude. And there is such a general want of forecast, that most of their misfortunes and hardships, in war and peace, come unexpectedly.[28]

After the Whig Party elected Indian fighter William Henry Harrison president in 1841, Schoolcraft lost his government appointment as agent. As a result the ethnographer returned back east, and worked as an independent scholar funded by Congress. His magnum opus—*Historical and Statistical Information Respecting . . . the Indian Tribes of the United States*—was published in six volumes between 1851 and 1857. It caught the attention of poet Henry Wadsworth Longfellow, who had spent his childhood summers among the primitive small lakes in Maine west of Portland near the New Hampshire border. He distilled the essence of Schoolcraft's work into *The Song of Hiawatha*, acknowledging it as his principal source and dedicating the book to Schoolcraft.[29] *Hiawatha* remains America's most famous epic poem as well as the most accessible account of Ojibwe subsistence for western ears.

Longfellow's protagonist was the pan-Algonquin spirit god Manabozho, whose name did not easily roll off the American tongue. So Longfellow appropriated the name of the "historic founder of the Iroquois confederacy" and changed it from Ayenwatha to Hiawatha. Into this merger, he incorporated the prairie Lakota in the form of

Minnehaha, the maiden destined to become Hiawatha's bride. "In a single bold stroke, Longfellow created out of the Ojibwe, a chimeric (good) Indian identity appropriate for mid-century America."[30] It bore little resemblance to the settled, powerful, antagonistic eastern tribes with whom the colonists had competed, nor to the bison hunters of the prairie at the other end of the glaciated fringe.

The social structure of the Ojibwe was dominated by extended familial networks called clans or kindreds, each with a home base and family camps in recognized areas. Society was loosely knit and highly flexible, being organized around gender, age, and specialized technological and spiritual skills, rather than hereditary castes. This organization was closely linked to the animals that shared the lake-forest ecosystem with them. Group affiliations were traced more through animal totems—analogous to a European coat of arms—than through genetic descent.

In 1885 the early Ojibwe historian William W. Warren recorded twenty-one totems, most of which were obligate denizens of lakes and wetlands.[31] Five were freshwater fish (pike, whitefish, catfish, sucker, sturgeon). There were eight birds (crane, loon, eagle, black duck, cormorant, goose, gull, hawk)—all but the hawk were lake-wetland creatures. Of the seven mammals (bear, marten, reindeer, wolf, lynx, moose, beaver), only two were obligate lake-wetland inhabitants, though all were well adapted to the habitat. Only the rattlesnake seems out of place today, but prior to the mid-nineteenth century, they were quite common on south-facing rock ledges of the Great Lakes.

The list of family totems reveals important food sources. Sturgeon, the largest of the totem fish, migrated seasonally up the principal rivers draining to the Great Lakes to spawn. The Ojibwe speared them in the shallow riffles of streams or trapped them in weirs, particularly upstream from the head of Lake Superior. Whitefish, found in most lakes, were captured in gill nets. Suckers were speared in shallows between ponds during spawning runs, whereas catfish were bottom feeders and easily took hook and line. Pikes were the only piscine predators on the list; largest of all was the muskellunge called

Moshkeenozha by the Ojibwe, which was admired more for its barracudalike power than as food.[32] Freshwater clams and mussels were also part of a diet largely dependent on aquatic food sources.

During my college days, I foraged the lake-forest ecosystem to supplement an uninspired grocery store diet. I harvested whitefish with gill nets and smelt with fine-meshed seines. In the spring I speared suckers in shallow narrows using something resembling a pitchfork with barbs. During the winter I speared northern pike through holes in the ice with a heavy iron spear. In the summer I dove for clams and crayfish, using my hands and a mesh bag to gather them from the bottom.[33] I fished with a hook year-round. My anchor starch came from the roots and rhizomes of cattails, supplemented by wild rice. I also foraged for greens and picked berries in the fall. The most challenging thing I ate was part of a porcupine.

Meat from terrestrial animals was also important to the Ojibwe diet, especially for obtaining fat. "The Chippewa were always a timber tribe and their principal native industry was the trapping of wild animals."[34] Of the seven animal totems that were neither fish nor fowl, four were regularly eaten for food. Moose, beaver, and bear seem to have been favored meats. Moose were hunted, whereas beaver and bear were trapped with various devices, including deadfall pits, nooses, and split logs. Additionally, beaver were killed with clubs, after their lodges were exposed by destruction of the dams or by chopping through winter ice to their dens. Bear were stabbed during hibernation or trapped.

But of all foods, wild rice was the most important. This beautiful, dark brown grain—also called foolish oats, nature's wheat, manoomin, good berry, and *Zizania palustris*—was the staff of life for the Ojibwe, and their Archaic predecessors as well.[35] Wild rice is the only cereal grain other than maize that was both native to the Americas and a source of sustenance for native populations before European contact. Planted corn fed the tribes of the eastern, southern, and western edge of the glaciated fringe. Unplanted rice fed those in the heartland. In northern Wisconsin, where beds of rice are especially common, they called it *omanomen, mano'min,* or *manoomin*. The people—*inini*—who ate it became the Menomenees, "people of the rice." Every tribe in the northern heartland considered it a gift of the gods.

Wild rice is an emergent aquatic grass with a substantially large seed head. It germinates best in soft muck, but must be able to root into something firmer at depth. These conditions are met in quiet standing water protected from wind and waves. The water must be at least several feet deep, warm enough to promote vigorous growth during a short growing season, and contain a moderately high nutrient level. It requires a late-summer climate dry enough so the grain heads don't rot before the September harvest. Most sites along the shoreline of the Great Lakes are either too cold, rough, or lean. But in their protected bays and in clusters of kettles between them, the conditions are ideal.

In 1660 the early French trader Pierre-Esprit Radisson left this journal account of eating wild rice with his native companions.

> Our songs being finished, we begin our teeth to work. We had a kind of rice much like oats . . . it grows in the water three or four feet deep . . . this is their food for the most part of winter, and they do dress it thus; for each man a handful of that they put in the pot, that swells so much that it can suffice a man.[36]

Jacques Marquette and Louis Jolliet watched them gather and prepare rice during the September of their famous 1673 voyage. When the seed head, or corn, is in an upright position, it is more likely to bend into the water during a wind or catch and hold rain, thereby enhancing rot. Native practice was thus to make two passes through the rice fields in latest summer or early fall by canoe. On the first pass they tied several heads together with a fiber of basswood and bent them over, which hastened ripening and kept the seed heads dry. On the second pass a person in the rear used one wooden stick to bend the stalks over the canoe and another to knock the grain from the stem.

The load of rice was then scooped from the canoe, winnowed on a tray of birch bark, and parched. High in protein and fiber, it had great dietary staying power, was rich in vitamins and minerals, and was completely portable. The Ojibwe and their antecedents survived on it directly for thousands of years. They survived on it indirectly because it helped feed the waterfowl on which they also depended.

I harvested wild rice as a teenager to earn extra spending money. The canoe gliding through the stems produced a staccato whisper, something like the brush of a snare drum. The wooden stick whacking the gathered heads of grain was a gentle percussion. Seeds sprinkled into the canoe. Water lapped and sloshed quietly against the hull, and drops fell from the paddle. The combination of these delicate sounds is forever etched in my memory, although I spent a lot more time harvesting wheat and barley on our Dakota homestead. There the sounds were on an industrial scale—the cacophonous roar of engines; the clattering mower; the whirring thrasher inside the center; the shaking, whining, and whishing of the separators, the auger, and the hopper, respectively. These louder sounds are harder for me to recall.

The Ojibwe remained firmly committed to wild rice as a dietary staple well into the twentieth century. In his 1902 "Wild-Rice Gatherers of the Upper Great Lakes," anthropologist Albert E. Jenks wrote, "The most princely vegetal gift which North America gave her people without toil was wild rice. They could almost defy nature's law that he who will not work shall not eat."[37]

Berries, especially cranberries, were also indispensable as flavoring agents, preservatives, and sources of vitamins. Pemmican was a staple, something like a combination of soft jerky, lard, and trail mix. Congealed animal fat, bits of meat, and berries were pounded together, molded or cut into cakes, dried, then wrapped to prevent their getting wet. In season, berries were a food source as well, one that could be gathered by children. Smoked fish were also portable food.

In his creation myth for indigenous native America, Longfellow wrote of maple sugar, known as *wendjidu'zinziba'kwid.*

> Unmolested worked the women,
> Made their sugar from the maple,
> Gathered wild rice in the meadows,
> Dressed the skins of deer and beaver.[38]

To make this delectable substance, maples were tapped in late winter, their sap gathered in birch-bark buckets, and the sap concentrated by one of many methods. Prior to European contact, the foraging Ojibwe did not have heavy soapstone and ceramic cooking containers in which sap could be boiled, suggesting that maple sugar was not important for subsistence. But small amounts of maple sugar were easily obtained by letting sap flow out onto sheets of birch bark to dry in the sun, by dipping hot stones into the sap as if dipping candles, or by letting it freeze and removing the ice. After the introduction of iron cooking vessels in the sixteenth century, maple sugar became an almost essential part of the diet. It was boiled down into granules and then stored for use as a confection, a sweetener for porridges and teas, and an agent of food preservation.

Food procurement by the Ojibwe reflects a heavy reliance on the mosaic of lakes, streams, small winding rivers, marshes, and bogs. Traversing such country in winter was fairly easy by snowshoe and dogsled, an invention originating even farther to the north. Getting around during the summer was much harder and required a watercraft technology especially adapted to local conditions. To the west, the Sioux used buffalo-skin round boats as ferries to cross their muddy rivers. Eastern tribes used wooden dugouts, which worked very well in shoreline and estuarine environments where no portages were required. Neither of these watercraft types would have worked in the kettle lake heartland. There the birch-bark canoe was indispensable. It was for the Ojibwe what horses were to the Plains Indians after the sixteenth-century reintroduction of this animal to the continent—a mode of transportation almost perfectly suited for the landscape.

One French word in particular helps explain why the birch-bark canoe was so successful. As early voyageurs adapted to Indian ways, they developed a unit of distance measurement called the *pose*, converted to "pause" by English traders. This was the distance between rest stops that someone could carry a canoe or some other heavy load being portaged. The native word was *opugidjiwunon*.[39] Given the variable soils over which portages occurred—sand, muck, marsh, bog, boulders—and the circuitous pathways associated with ice stagnation topography, a functional distance given in pauses was far more revealing for describing a journey than a linear distance in miles. The lighter the canoe, the longer the pause, the faster the travel, and the less time it took to move about on seasonal rounds. Lightness in weight, especially when portaging over soggy, if not quaking, ground, became the most important criterion for a watercraft beyond the fact that it floated.

Throughout postglacial time, many shallow lakes and the entrances to deeper ones were being progressively filled with bog and marsh. Through them meandered narrow, sluggish streams often just a few feet wide. Under such conditions, neither the horizontally extended paddles of a kayak or the oars of a rowboat would have worked very well. And, given the ubiquity of pebbly sand in the Ojibwe homeland, swift-flowing reaches of streams were typically very shallow. A flat-bottomed canoe without a keel was the ideal answer to this depth constraint. A keel would have been useful to help stabilize a canoe on larger lakes, but its added weight would have shortened pauses and would have interfered with the chronic need to turn sharply.

Birch bark is only one of several critical ingredients for a canoe. Cedar was indispensable because it split into strong, thin, flexible bands to create the frame. The fibers from spruce roots were used to tie the bark strips together and to the frame. The resin of conifers—spruce, fir, and pine—was used to seal the seams. The size of canoes was adjusted to the weight of the load and the size of the water body. Huge "Montreal" canoes about thirty-five feet long were used for the annual trek to the fur rendezvous. They had to be large enough to carry a heavy load of furs and to survive the storms on the Great Lakes. Foraging canoes for the marsh-lake system, however, were typically only twelve to fourteen feet long, barely big enough to carry two people and bring back the carcass of a moose or a bear.

Archaic, Woodland, and Contact period natives of the lake-forest country would have starved without their canoes just as surely as the southern tribes would have starved without their hoes and digging implements. The same was true for the voyageurs, who quickly adapted to the "wandering forester" lifestyle.

## 5.

# The Fur Trade, the Northwest Passage, and the Source of the Mississippi

CLIMATE CHANGE WAS DIRECTLY RESPONSIBLE for the stabilization of the lake-forest ecosystem about ten thousand years ago and the rise of the Archaic foraging pattern of human subsistence. It was also indirectly responsible for the demise of this human ecosystem, largely through the destructive impact of the fur trade and its legacy.

The Little Ice Age struck northern Europe with a vengeance between the fourteenth and nineteenth centuries. Mountain glaciers advanced in the Alps and crushed buildings. Estuaries froze, stunting the economy of port cities and fishing villages. Shorter growing seasons led to food scarcities, plagues, and feudal wars, beginning with the "calamitous 14th century" and ending with the bitter cold winters of the 19th century.[1]

There was a modest drop in the average air temperature and an increase in dampness over Europe. Winter cold penetrated stone castles and peasant hovels alike. Flaming fireplaces and smoldering cooking fires were burning what was left of the native forests before the age of King Coal. The best solution to the combination of persistent cold and a fuel shortage was to wear furs.

Thick, supple animal skins warmed the nobility of England, Scandinavia, Germany, and Russia, while the peasants wore what they could afford. As the centuries of the Little Ice Age dragged on, the supply of Old World furs steadily diminished, and traders sailed across the Atlantic to procure them from the New World. Demand for beaver pelts remained high through the middle of the nineteenth century owing to a haberdashery fad for stovepipe hats in the era of Jane Austen, Charles Dickens, and Abraham Lincoln. And the best beaver pelts of all came from the northern lakes and forests of the glaciated fringe and the adjacent Canadian Shield.

Pelts there were of a "deeper color and better luster" than those from the warmer creeks south of the glacial limit or from the drier plains to the west. Northern beavers had evolved their extra-thick and well-oiled fur coats in response to the double-edged environmental sword of living an aquatic life in a place with a bitterly cold winter climate.[2] This evolutionary advantage for the beaver turned out to be its downfall during the fur trade. Like a powerful magnet, the supple skins of this aquatic mammal drew courageous Europeans deep into the heart of North America. "The beaver," wrote author William Least Heat Moon, "almost as much as the horse, helped shape the course of early American history."[3]

An entry from a nineteenth-century ledger from a trading post on the Minnesota-Ontario border at Grand Portage helps us understand the magnitude of the fur trade.[4] In a single year from a single place, a single trader sent 1,621 beavers, 1,456 martens, 12,470 muskrats, 862 wolves, 509 foxes, 507 minks, 469 moose, 332 fishers, 214 otters, 152 raccoons, 125 black bears, 49 brown bears, 45 wolverines, and 4 grizzly bears. All of these north-woods creatures were either dependent upon or adapted to the shorelines of small lakes and ponds, generally rock-carved ones to the north and kettles to the south.

During the early seventeenth century, there was intense competition between French, Dutch, English, Swedish, and American enterprises. In the end, the English prevailed. Their pioneer was Captain John Smith, of Jamestown, Virginia, fame. In 1614, six years before the Pilgrims anchored at Truro, he was trading trifles for luxurious pelts along the coast of what he called Norumbega after a fabled Indian city that never existed. He wrote:

> Our plot was there to take whales and make trials of a mine of gold and copper. If those failed, fish and furs was then our refuge . . .

ranging the coast in a small boat, [from the Indians] we got for trifles near 1,100 beaver skins, 100 martens, and near as many otters.[5]

Whaling didn't pan out. Nor did prospecting for minerals. Fur became their "refuge" from the financial calamity of returning home empty-handed to disappointed investors. His fur trade, however, was not a fair trade, because he exchanged "trifles" like those given away at carnivals today for perhaps the hottest market commodity of the era.

Captain Smith inaugurated an English fur monopoly along the Atlantic coast that would extend from southeastern Massachusetts to Newfoundland. By 1628 the Plymouth Colony—usually thought of as a pious people seeking a haven from persecution—was getting most of its income from beaver pelts. In another English enterprise more than a quarter of a million beaver pelts were sent to London from the Connecticut River Valley, where William Pynchon ran a legalized monopoly.[6] The English fur trade in New England collapsed for lack of pelts by 1670. In response, King Charles II chartered an operation out of York Factory, on the southwest side of Hudson Bay. With so little else going on away from the Atlantic, the Hudson's Bay Company became the de facto government for the hard-rock interior of Canada for more than a century. The presence of this company secured a British dominion that would eventually translate into Canada.

Other colonial powers seeking a piece of the action were eclipsed by the British. In 1624 the Dutch West India Company was monopolizing the fur trade along the southern coast of New England and up the Hudson River. Beaver remained the dominant export from New Amsterdam until the end of the seventeenth century, when the Dutch presence largely disappeared. A Swedish colony also developed along the Delaware River in western New Jersey at that time. It was composed mostly of rural settlers from Finland—then part of Sweden—whose mixed economy of hunting, fishing, and small-scale farming closely resembled that of the eastern Woodland tribes. The Swedes were squeezed out by the Dutch in 1655, and the Dutch by the British in 1664. By that time the most significant visual icon of the American settlement experience, the log cabin, had been built on American soil, and it remains an important part of twenty-first-century lake culture.

French fur traders and explorers were the only ones to penetrate deeply into the glaciated fringe, beyond the limit of ship navigation. Jacques Cartier made his first voyage up the Gulf of St. Lawrence in 1534. Within a century, and long before the English got serious about trading furs, the French were heavily invested in the game and had moved far inland along the Great Lakes. Every summer a flotilla of enormous birch-bark canoes carried bales of furs eastward to Montreal for the annual fortnight-long fur rendezvous, equal parts trade show, mercantile exchange, and feast.[7] By the end of the seventeenth century, the French had penetrated all the way to the prairies flanking the Missouri River, terrain that would later be claimed as part of Louisiana.

Although the North American fur trade lasted nearly three centuries, on a local scale it was a boom-bust business. The basic plan was to "take what there is, take all of it, and take it as fast as you can, and let tomorrow's people handle tomorrow's problem."[8] For example, Alexander Henry of the North West Company, based out of Montreal, established a trading post in Pembina, North Dakota, planning to export pelts from what was then virgin territory near the north-draining Red and Assiniboine rivers. Business began in 1801, picked up quickly, and peaked in the 1804–1805 season, when 2,736 beaver pelts were traded. By the 1807–1808 season, however, the yield had crashed to 696 pelts.[9] The seven-year interval between boom and bust was comparable to that of the California Gold Rush or later rushes to the Klondike in the Yukon Territory.

The only thing that kept the fur trade going was relentless expansion into new territory. In the wake of this expansion was the local near extinction of fur-bearing animals, whose populations crashed to the point that they weren't worth going after. Another near extinction was of a native culture that had been a stable part of the ecosystem for millennia. Although French, English, Dutch, Swedish, and Americans adopted native languages and woodcraft, took Indian wives, created a mixed race of children, and forged North America's first political alliances, Europeans ultimately heralded a death knell for the foraging way of life. A Michigan historian put it this way: "The Indian only seemed to be living in the Stone Age. He really was a part of eighteenth-century Europe; he was working for the white man every day of his life."[10]

Natives employed by the fur trade were, in effect, rural servants of a European commodity exchange. As with sweatshop labor today, they did most of the dirty work—trapping, killing, skinning, tanning, and hauling—merely to make ends meet. No longer were their main activities seasonal rounds of fishing, hunting, and gathering, followed by the winter slowdown when culturally rich lodge stories were told. No longer were full bellies, healthy children, and peace with neighboring clans the goal. Now the goal was to obtain cash for market exchange. "The beaver," one Indian remarked, "makes kettles, hatchets, swords, knives, bread; and, in short, it makes everything."[11] Once in possession of European goods, Indian men and their families became dependent on cash for replacement tools, gun parts, and bullets, and for flour, sugar, tea, and alcohol. This forced the native trappers to intensify local efforts and spend more time farther away. With the men being absent more, the women, children, and elderly abandoned traditional village sites and moved closer to trading posts for their own protection. There they experienced constant, often coercive pressure to adopt European ways.[12]

Hostilities between territorial native groups increased as a result of competition for imported goods, increased firepower, and the effects of alcohol. As a result, Indian tribes experienced an east-to-west domino effect of territorial usurpation because steel weapons and firearms were acquired by eastern tribes first. The Ojibwe, headquartered near Sault Sainte Marie for at least several centuries, lived at the midpoint of this cascade. As a result, their tongue became the court language, the lingua franca of the day, even among the French.

The Ojibwe were pushed west by groups displaced by the Iroquois. In turn the Ojibwe pushed the Santee Sioux from eastern Minnesota to the edge of the prairie, which displaced resident Lakota tribes. Territorial skirmishes between the Ojibwe and the Sioux began as early as 1642 and continued for nearly 250 years.

The fur trade accounted for only half the economic motive to penetrate the continental interior. The other half was to discover a passage to the Pacific. Christopher Columbus didn't so much discover America for Spain as bump into it inconveniently on his way to the Orient. Later Spanish expeditions concentrated their search for a passage to the south, especially the ruinous Hernando de Soto. English expeditions concentrated their search to the north, failing on multiple tries between the fifteenth-century voyages of John Cabot and the fatal nineteenth-century voyage of Sir John Franklin. The Dutch tried the Hudson River by employing Henry Hudson, getting no farther than Albany, New York. The French search via the Gulf of St. Lawrence ended quickly. In 1535, from a hill above what is now Montreal, Jacques Cartier saw, "to his dismay," a series of rapids, "a sault of water," that he knew would stop all ocean-going ships.[13] From there, westbound traffic would be by canoe.

Defeated by the rapids, Cartier returned downstream and overwintered near the present site of Quebec, where he mingled with the natives. His stay inaugurated a material and cultural exchange that would create a far-traveled group known as the *coururs de bois*, more popularly known as the voyageurs. This would be an era of muzzle-loading muskets, log forts, and Jesuit missionaries. Etienne Brulé, a protégé of Samuel de Champlain, was the earliest voyageur of note. By 1602 Brulé had reached Lake Superior, the westernmost, grandest, coldest, and clearest of the Great Lakes, one year before Champlain, his mentor, founded the city of Quebec. By 1613 Champlain had discovered a shortcut to the western Great Lakes via the Ottawa River and a portage to Lake Huron's Georgian Bay, which is both large enough and enclosed enough to qualify as a Great Lake in its own right. The final link in western penetration was completed in

1659, when Pierre-Esprit Radisson and Médard Chouart des Gro-seilliers discovered the Grand Portage between Lake Superior and more distant lakes to the northwest, the largest of which is Lake Winnipeg in Manitoba.[14] This important discovery connected the voyageur canoe routes of the St. Lawrence–Great Lakes system with those of western and northern territories, specifically the enormous Hudson Bay and Missouri River drainages.

The "chain of lakes" used by the voyageurs follows what geologists call an "arc of exhumation" parallel to the southern edge of the Cana-dian Shield. There the ancient, harder rocks of the shield have been exhumed by the glacial erosion of the younger, softer rocks of the platform.[15] Erosion was especially intense, due to a thermal transi-tion at the base of the ice sheet toward freezing-on conditions and because the bedrock is slivered by geological faults into thousands of fragments. The end result was a concentration of countless bedrock lakes aligned along the grain of rock resistance and connected to each other by deep meltwater channels. After the Grand Portage, only eight miles inland from Lake Superior, the remaining portages for the next five hundred miles were typically short, took place on solid rock, and went around rapids and low waterfalls. The center-piece of this bedrock canoe highway is preserved today as Voyageur National Park, located within the Boundary Waters Canoe Wilder-ness, shared between northeastern Minnesota and southwestern On-tario. It remains to this day a rugged, hard-rock wilderness that can be most easily crossed by canoe, kayak, or light rowing craft.

GREEN BAY, located on the western shore of Lake Michigan, is the oldest city in Wisconsin. It is also one of the most important jumping-off places in American history. There, in 1634, an intrepid and misguided explorer from Normandy named Jean Nicollet de Belleborne stepped out of a canoe fully prepared to greet the man-darins of China. Instead he encountered a rich and exotic native cul-ture called the Winnebago, descended not from the more familiar

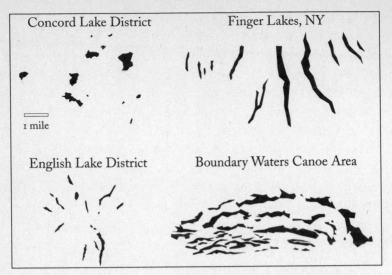

Concord Lake District      Finger Lakes, NY

1 mile

English Lake District      Boundary Waters Canoe Area

*The Concord lake district (at 10X magnification) compared to others.*

Algonquin tribes of the east but from the unfamiliar Sioux tribes of the Missouri and Mississippi river drainages. (Winnebago was a common name used to describe the historic tribes of this area, the Ho-Chunk and Oneida nations.) The difference in native customs must have been a clue to Nicollet that he was on the edge of something special. But this was no Northwest Passage to the Orient. Instead it was a southwest passage to a vast expanse of land later claimed as the Louisiana Territory.

The Winnebago tribe took its name from a lake to the south, a shrunken remnant of an enormous former glacial lake in the same lowland basin occupied by Lake Michigan's Green Bay. Though the third-largest lake wholly within the United States, Lake Winnebago is remarkably shallow, especially on its eastern side, and with an average depth of fifteen feet. It is fed by streams draining northward from the richer soils of deciduous forests in southern Wisconsin toward the leaner conditions of the northern pines. The higher nutrient level of the lake water, the enormous surface area of the lake relative to its depth, and its muddy margin made the lake warmer and more fertile than normal, and therefore unusually rich in fish, game, and plant foods. Its outlet through the Fox River enriched Green Bay as well, causing intermittent natural algal blooms during the unpolluted seventeenth century that tinted the water green and gave the bay its name.

After living among the tribe for the winter, Nicollet returned to Quebec. Within a year, his sponsor, Samuel de Champlain, died, beginning a forty-year-long hiatus of French exploration to the southwest. But in 1673, fourteen years after the bedrock routes of the voyageurs to the northwest were completed, Father Jacques Marquette and Louis Jolliet followed Nicollet's route, paddling southwest across Lake Michigan to Green Bay. From that point, they canoed up the north-flowing Fox River into Lake Winnebago and continued southwesterly up its main-inlet stream. Eventually, the adjacent hills began to close in as Marquette and Jolliet crossed a moraine, through which had been carved a broad, flat-bottomed meltwater channel. After a swampy portage where the Fox River flows both ways, they entered the south-flowing Wisconsin River and continued downstream to its mouth at Prairie du Chien, Wisconsin, opposite northeast Iowa.[16] There they encountered the "Father of Waters," variously known by the natives as the Mechassipi, Micissipi, Meschasipi, Mezzisipi, and Mississippi. Marquette and Jolliet descended to the Arkansas River, at which point they turned around, fearing hostile natives and a rumored Spanish presence. Though they did not reach its mouth at the Gulf of Mexico, they were able to confirm that the great, south-flowing river discovered more than a century earlier by Hernando de Soto, in 1540, extended northward to the midcontinent.

Following the advice of their Indian guides, Marquette and Jolliet returned to Lake Michigan by a shorter, easier route up the Illinois River, which took them toward the main body of the lake, rather than its principal bay. Along this route the resident Indians were called the Illini, namesake for the state of Illinois when given a Francophone spin. At the head of the Illinois River, they followed another large meltwater channel cut across a kettle moraine similar to the one in

Wisconsin. At the northern end of this channel was a wetland draining northward into the Chicago River, which flows into Lake Michigan. This wetland sits in the saddle of the American landscape, the
place where the upward arch between the St. Lawrence and Mississippi River drainages lies near the downward trough formed by the
convergence of the Ohio and Missouri River drainages. It's no acci-
dent that Chicago, the largest city in the Midwest, was built where
protected lakeshore lies nearest to the center of the saddle.

Seven years later, in 1680, Father Louis Hennepin followed the reverse route up the Chicago River from Lake Michigan and down the
Illinois River to the Mississippi. But instead of floating downstream
in the direction of Marquette and Jolliet, he paddled upstream toward the unexplored north country. Captured by the Santee Sioux
tribe, his party was taken as prisoners and brought north near what
would later become the Twin Cities of Minneapolis and Saint Paul
in Minnesota. There he saw and named Saint Anthony Falls, the
only major waterfall on the entire Mississippi, one that would later be
used to power the flour mills for immigrant-grown wheat and to cut
logs for the timber trade. Father Hennepin was rescued by Daniel
Greysolon Sieur du Lhut, who, during the previous year, had
reached the western tip of Lake Superior at Duluth, as the city was
later named for him. Du Lhut reached the Mississippi by paddling
up the north-flowing Brule River from Lake Superior, portaging
across a swampy meltwater channel cut through a moraine, and floating down the south-flowing St. Croix River.

The most far-traveled French explorer of the era was René-Robert
Cavelier, Sieur de La Salle. Under orders from King Louis XIV, in
1682 he paddled the Chicago and Illinois rivers route to the Mississippi before floating all the way down its brackish, alligator-infested
delta. He claimed the entire watershed of the Mississippi—especially
to the west—for France, naming it Louisiana in honor of the king
sponsoring the trip.

All of these great historic journeys for New France—by La Salle,
Hennepin, Marquette, Jolliet, Nicollet, and Du Lhut—began on the

same path followed by the fur traders, the oceanic surfaces of the Great Lakes. What the explorers did not know then was that even after leaving the hard-rock shores of the modern Great Lakes, they remained within the muddy shores of greatly expanded versions of the Great Lakes that had been dammed to the north by the retreating ice front. These coalesced ice-front lakes wrapped around the retreating Laurentide margin from near Montreal on the east all the way to Great Slave Lake in Canada's Northwest Territory. The largest was Glacial Lake Agassiz, which lay west of Wisconsin and was the largest lake ever known to have existed on Earth.[17] To the north it was deep and had a fairly straight border because the crust had been depressed by the weight of the ice; there the glacier calved icebergs into gray water. To the south the lake was muddy, shallow, and had a crenulated margin. Lake Agassiz and lesser (but still impressive) glacial lakes to the east were forced to drain southward via the Missouri-Mississippi-Ohio system because the ice sheet still blocked the St. Lawrence and Hudson Bay drainages. Among these were Glacial Lakes Aitkin, Wisconsin, Oshkosh, Wauponsee, Maumee, Warren, Whittlesey, and Iroquois.

Ice-front lakes seldom drained completely because their bottoms were lined with nearly horizontal sheets of clay and silt, and the crust beneath that was tilted southward toward moraines. Upper Red Lake, Lower Red Lake, and Mille Lacs Lake in Minnesota are remnants of ancient ice-front lakes that catch and hold modern rainfall. They look like oversized kettles, but have smoother shores because the muddier materials were more easily worked by waves into sweeping curves. The Red River Valley between Minnesota and North Dakota is a filled remnant of Lake Agassiz. There the lake was so shallow that it became a wide, swampy floodplain. Today the river is a tightly meandering stream flowing toward Hudson Bay with a gradient so slight that it reinvents itself as a lake during floods. Near the southern limit of Lake Agassiz lies the site of present-day Fargo, North Dakota. John Steinbeck's description of Fargo as "blizzard-riven, heat-blasted, [and] dust-raddled" is a legacy of its ice-age setting. Winds sweep uninterrupted across the vast flat expanse of the former lakebed, which is heavily cultivated because it is some of the richest mud soil in the nation.[18]

Farther to the south and east, the flat bottom of an expanded Lake Michigan gave rise to the great marshes and swamps near Chicago (Glacial Lake Glenwood), the crane marshes so evocatively described by Aldo Leopold (Glacial Lake Wisconsin), and the wonderful cropland surrounding Saginaw Bay, Michigan (Glacial Lake Nipissing), and Detroit (Glacial Lake Warren). Similarly, the Kankakee Swamps along the Maumee River in Ohio result from an expanded version of Lake Erie, and those below the Adirondacks, an expanded version of Lake Ontario called Glacial Lake Iroquois.

At the southern limit of every single one of these ice-front lakes was a meltwater spillway that controlled its level. The most colossal was at the south end of Lake Agassiz parallel to the border between South Dakota and Minnesota. There the modern-day, ribbon-shaped Lake Traverse and Big Stone Lake occupy the bottom of a deep, low-gradient meltwater trench up to five miles wide and cut to a depth of 250 feet. This was the canoe portage used by the voyageurs between Hudson River and Mississippi drainages. All of the other north-south portages used by the explorers of Louisiana were also former meltwater spillways cut across moraines, through which flowed torrential overflows. Poet Carl Sandburg imagined the noise: "Here the water went down, the icebergs slid with gravel, the gaps and the valleys hissed . . . "[19]

Only when the Laurentide Ice Sheet withdrew from the eastern Great Lakes about twelve thousand years ago were the southern ice-front lakes free to drain through the St. Lawrence to the Atlantic in a great iceberg flood that changed the global climate. Their spillways were abandoned, each leaving behind a subdued box canyon with a negligible gradient, especially to the north where the crust was being tilted upward. Geologists refer to these valleys as coulees, from the French word for "flow."[20]

By the time the French explorers arrived, the coulees had become quiet marshes, from which small streams issued in both directions. In some cases, portages weren't even necessary—one could simply paddle up from a Great Lake, through a marsh, and then down tributaries to the Mississippi. The swampy portages that first opened up the Louisiana Territory lie at the aptly named city of Portage, Wisconsin, birthplace of the great historian of the west Frederick Jackson Turner. Knowledge of that connection may have inspired his thinking.

A THIRD TYPE OF CANOE ROUTE would be needed to explore the low plateaus standing above and between the expanded glacial versions of the Great Lakes. To reach the headwaters of large regional rivers—the Manistee, Au Sable, and Muskegon rivers in Michigan; the St. Croix, Chippewa, and Wisconsin rivers in Wisconsin; and the St. Louis, Minnesota, and Red rivers in Minnesota—required navigating through a maze of kettle lakes, marshes, and winding streams within ice stagnation terrain. Most noteworthy was the elusive and frustrating search for the source of the Mississippi River, historically the most important river in the United States.

When La Salle claimed the Louisiana Territory in 1682, he knew the Mississippi extended from the Gulf of Mexico to somewhere north of Saint Anthony Falls, Minnesota, which Hennepin had seen two years earlier. But its path above that point, presumed to extend as far north as the fiftieth parallel, was anybody's guess.[21] In 1688 the French cartographer Jean Baptiste Louis Franquelin guessed wrong when drawing the definitive map of Louisiana, on which later negotiations were based. His mistake was to extrapolate the Mississippi as a straight line between two well-charted points, Saint Anthony Falls and Lake of the Woods, which by that time was well known to the fur traders. The reality is that the Mississippi River doesn't reach anywhere near Lake of the Woods, which intersects the present U.S.-Canadian border at the forty-ninth parallel.

Instead, it traces out a giant question mark through the "deranged topography" of north-central Minnesota. The straight part of the question mark lies below the center of the state near Brainerd. No

*The Mississippi River at its source.*

problem there. From there, however, it curves northeast to Jacobsen, northwest to the top of the curve at Bemidji, southwest to Coffee Pot Landing in Clearwater County, then finally southeast to the small kettles and bogs beyond Lake Itasca. As with any question mark, the area in the center is simultaneously west, south, and east of the curve, in this case the trace of the Mississippi River. This means that the boundary of Louisiana as originally conceived was impossible to define.

Between the voyageurs' chain of aligned bedrock lakes to the north and the coulees used by the explorers of Louisiana to the south lay nearly two hundred miles of fairly flat, lake-speckled landscape. Traversing this country required paddling across countless kettle lakes with multiple bays, navigating the tortuously winding rivers between them, crossing quivering savannahs and bogs, and portaging up and over countless sandy moraine ridges. In short, the landscape mosaic that proved so ideal for Native American foraging helped delay the discovery of the headwaters of the Mississippi for more than two centuries. This is a story of international significance because it involved back-and-forth negotiations between five nations: France, Spain, England, the United States, and what would become Canada.

Establishing the first part of the international boundary was easy. It lay between what is now northeastern Minnesota and Ontario along the well-known bedrock canoe route up from Lake Superior to the Lake of the Woods.[22] From there, the boundary was drawn arbitrarily westward along the forty-ninth parallel. All of the continental United States (including northern Maine) now lies south of this line, except for the Northwest Angle, a burr of land sticking up at the triple point where Minnesota, Ontario, and Manitoba meet. This burr is an artifact of mistaken assumptions about the source of the Mississippi and subsequent cartographic reshufflings.

France transferred the Louisiana Territory to Spain in 1762. This divided present-day Minnesota into a Spanish west, an unknown north, and an eastern section being contested by France and England. With the end of the French and Indian War in 1763, the British assumed control of all lands east of the Mississippi; land west of it remained Spanish. Within twenty years, the territory near the eastern headwaters became part of the fledgling United States of America, taken from Britain at the end of the Revolutionary War in 1783. But the British stayed there anyway, possessing it illegally until well after the War of 1812. The western headwaters territory was transferred from Spain back to France in 1800, which sold it to the United States of America as part of the Louisiana Purchase in 1803. Throughout it all nobody but the natives knew that the lands being traded lay inside the enormous question mark loop.

To help solve the growing tension regarding the boundary issues, Alexander MacKenzie of the North West Company appointed David Thompson to find the Mississippi River headwaters relative to the forty-ninth parallel. Setting out in 1797, Thompson combined his knowledge of astronomy with a telescope, sextant, and compass to prove that the Mississippi headwaters lay well to the south of the forty-ninth parallel. His mistake was to designate the Turtle River northwest of Bemidji as the source. His remapping, however, did shrink the northern limit of Louisiana by more than a hundred miles.

Had the true headwaters of the Mississippi been known when the sale of the Louisiana Territory was being negotiated, the final boundary between what is now Canada and the United States might have been drawn either at the northernmost point of the Mississippi, at Bemidji, or at its source, at Lake Itasca, well to the south. These points are at 47.30 and 47.12 degrees north latitude, respectively. Had the boundary been drawn at Itasca, Canada would now include northwestern Minnesota, most of North Dakota, about half of Montana, the Idaho Panhandle, and most of Washington State, including all of Puget Sound. Only after the Louisiana Purchase was securely signed and Lewis and Clark were on their way to the Pacific did President Thomas Jefferson wisely commission a definitive search for the Mississippi headwaters. What he did not expect was that the source would remain in limbo for twenty-nine more years, until a time well after his own death.[23]

The first attempt was an 1805 military expedition led by the great explorer Lieutenant Zebulon Pike, of Colorado's Pike's Peak fame. The expedition set off from the lower Mississippi River with a seventy-foot keelboat and twenty soldiers in early August. Delayed by the difficult travel, they got trapped by the northern winter and nearly froze to death. The Pike expedition designated crazy-shaped Leech Lake as the main source of the Mississippi. There, and with great ceremony, they shot down the British flag, which had been flying illegally at a fort on American territory for more than two decades. Making a side trip with smaller canoes, the Pike expedition got as far north as Upper Red Cedar Lake (now Cass Lake), where Pike was cordially treated to a feast of boiled moose head and roast beaver. His party then returned, leaving the source of the Mississippi undecided.

A second military expedition was carried out in 1819 by Stephen Long, a navigation surveyor and topographical engineer. He failed as well, defeated by the cold and kettle-crazed terrain.

Lewis Cass, governor of what was then called the Michigan Territory, carried out a third attempt in 1820. He set out with James Doty (first governor of the Wisconsin Territory) and Henry Rowe Schoolcraft, at the time a young mineralogist who had been recruited by the U.S. secretary of war to look for mineral wealth. The Cass expedition got no farther than did Lieutenant Pike, but officially proclaimed Cass Lake as the source of the Mississippi.

A final failed historical attempt came in 1823 when a romantic Italian count and political exile named Giacomo Constantino Beltrami set out on the quixotic search to locate the headwaters for personal glory. After a series of wandering and dangerous misadventures, he published a book pronouncing that the source lay far to the west at Lake Julia, which was way off.[24] All he confirmed was how easy it was to get lost up there.

Schoolcraft did not believe the 1820 party had succeeded in discovering the Mississippi headwaters; therefore, he mounted his own expedition in 1832, on the pretense of investigating the military strength of the Indians and their susceptibility to contagious disease, collecting scientific specimens, and ameliorating tensions between the Sioux and the Ojibwe, who were engaged in a war of skirmishes. Accompanying Schoolcraft were George Johnston, his Ojibwe brother-in-law, who served as interpreter and baggage master; Lieutenant James Allen and his soldiers; Dr. Douglass Houghton, the party medical scientist; and the Reverend William Thurston Boutwell, whose plan was to save a few Indian souls. During their journey, the party reported some unusual sights such as "voracious, long-billed dyspeptic mosquitos," the "common pigeon which extends its migrations over the continent," bogs and marshes they called "shaking savannahs," and a ceremonial scalp dance at Cass Lake in which the Ojibwe were celebrating a recent raid on the Sioux.[25]

The Schoolcraft party succeeded in reaching the source of the Mississippi where others had failed, in part because they were lucky. As they left Lake Superior to begin their search, they had the good fortune to stumble onto a small party of Ojibwe who knew precisely where the headwaters were. One of them was related to a chief named Oza Windib, who would become their guide, assign two of his tribe to accompany the men, plan their journey on a paper map, loan them the small hunting canoes needed to negotiate the small streams, and lead each portage. To Schoolcraft's surprise, Oza Windib guided them away from the main stem of the Mississippi at a point just south of

Bemidji, taking them up a smaller, unnamed, north-flowing stream now called the Schoolcraft River.

The final link to Itasca was a six-mile portage over a series of forested moraine ridges and soft marshes. Schoolcraft described the discovery moment: "What had been long sought, at last appeared suddenly. On turning out of a thicket, into a small weedy opening, the cheering sight of a transparent body of water burst upon our view. It was Itasca Lake—the source of the Mississippi."[26] The name Itasca was made up, having been derived from the Latin words *veritas* (true) and *caput* (head). He imposed this name on a body of water that had been known as Elk Lake, the English translation for Omushkos, the original native name, and as Lac la Biche by French trappers.[27] Having made their discovery, the exploration party spent only four hours on the lake, apparently because Schoolcraft "did not want to be criticized for an unauthorized excursion."[28] They floated downstream through the lake outlet and into unexplored territory until they reached the fork they had taken on their way up, what is now

*Itasca Lake, by Seth Eastman (1808-1875), shows
Henry Rowe Schoolcraft's discovery.*

called the Schoolcraft River. Only then did they know for certain that the stream they had floated down from Itasca was, in fact, the Mississippi, rather than some other river.

Under normal conditions, finding the head of an important river is simply a matter of working your way upstream, taking the strongest branch, and repeating the process until there are no more branches. This method didn't work in the ice stagnation terrain of central Minnesota, where sluggish marshy tributaries enter large lakes in seemingly random locations. There was simply no way for the explorers to know where the tributaries were until circumnavigating each lake perimeter. And even then it would have been hard for them to determine which incoming stream was the largest because the inflows were dissipated by marshes. The Mississippi may be the only globally significant river whose source was reached by an overland portage, rather than by working upstream through the drainage network. It may also be the only one whose discovery was made during an unauthorized side trip from a mission funded for other reasons.


DURING THE INITIAL CONTACT PHASE, Indians were seen by Europeans as an exotic people and potential converts to Christianity, later as indispensable field-workers for the fur trade, then as proxy soldiers in battles between colonial powers. But after the curiosity about native customs wore off, after missionaries realized that the Indian mind-set was stubbornly resistant to Western religion, after the fur trade went bust, and after their mercenary services were no longer necessary, the Indians were of no practical value to the new arrivals.[29] With the natives now part of the market economy, and with no obvious source of income, the most valuable thing they had left to sell was real estate. Many of the land swindles to come were preconditioned by this simple economic fact.

The extinction of the lake-forest foraging culture was bracketed by military genocides. The Pequot Massacre at Mystic, Connecticut, in 1637 was a deliberate attempt to render the tribe extinct. Land taking in the eighteenth century culminated in native reprisals and colonial counterreprisals associated with King Phillip's War (1675–76), the most costly campaign in U.S. history measured in proportion to the population. By the early nineteenth century, natives east of the Mississippi were seen as "varmints," second-class citizens at best.[30] Andrew Jackson and William Henry Harrison, in 1829 and 1841, respectively, won their presidential elections in part because of their "take-no-prisoners" images as successful Indian fighters.

As Manifest Destiny played itself out, the Indians were shunted to reservations where the land was poor, kept quiet with annuities, and taught American ways. Thoreau witnessed the beginning of the end of the foraging way of life during his trip to Minnesota, which he took for medical reasons. There, on the edge of the prairie, he met hordes of poverty-stricken Indians reduced to begging for handouts and so enraged by U.S. policy that they would wage war on settlers the following year during the Great Sioux Uprising of August 1862 led by Little Crow, who Thoreau had a chance to meet. Thoreau's final deathbed utterance, "Indian," may have been made with this despair in mind. The ignominious end of the Indian Wars finally came in 1890 when hundreds of Sioux women, children, and elders were slaughtered by the U.S. Army at Wounded Knee, South Dakota.[31]

"It has been said that Indians began with everything—the whole country was once theirs—and ended with close to nothing."[32] According to the Ojibwe writer Winona LaDuke, by the close of the nineteenth century, the government of the United States had "signed 371 treaties with Native people and made some 720 related land seizures on Native territory."[33] Native culture did not go extinct. But the foraging human ecosystem from which it emerged did disappear.

# 6.

# Kettles and Early America

KETTLE LAKES AND PONDS PROVIDED special resources for every phase of America's history.[1] During the first Pilgrim foray in 1620, it was potable water they sought. Kettles were also involved in their later decision to leave the outer Cape at Truro and resettle across the bay in Plymouth: the ponds were simply too high above the shore.

The next resource provided by kettles was hay for English livestock, especially cattle. Throughout New England, freshwater marshes provided a natural hay of superior quality to that of tidal salt marshes. Called "meadow" by the colonial English, it grew most luxuriously along the low floodplains of seasonally inundated large rivers like the Connecticut, Merrimack, and Saco. Next in importance to large-river marshes were those on the drained beds of former glacial lakes, especially those that had been kept wet by northward tilting caused by the relaxation of the earth's crust following the removal of the ice. Concord, Massachusetts, the oldest inland settlement within the glaciated fringe (1635), was nucleated by one such marsh. Historian Brian Donahue wrote that it "lay at the heart of the [mixed husbandry] system," because it generated not only milk and meat, but also the manure that kept tilled fields fertile.[2] Unfortunately, this type of lakebed marsh is quite restricted in New England.

Away from large river and lakebed marshes, natural hay was found most abundantly on the margins of shallow kettles that rose and fell each year. In such places, referred to as drawdown ponds by hydrologists and erroneously as coastal plain ponds by botanists, the groundwater table rises each spring, leaving a fertile coating of muck and preventing tree growth, but falls in summer to allow the growth of lush grass. Such kettles are widely distributed throughout New England and upstate New York, and likely nucleated many eighteenth- and nineteenth-century farms because keeping the cattle alive for those first few years of forest clearing was an essential first step. So valuable was freshwater hay from kettles that some were protected by dikes to prevent late-summer flooding by thunderstorms.

Farms of the minutemen and their Yankee descendants usually had a mixture of landforms and soils spread out over a largely pastoral landscape: Lodgment till on smooth hills was used for pasture; stonier lands were used for woodlots, orchards, and maple sugaring; and areas of loamy sand and gravel were used for cultivation. Kettles, when present, were usually located in very dry soils, and used as sites of special resources. In addition to gathering hay, settlers watered their stock, cut cedar for fence posts, picked medicinal herbs, harvested blueberries and cranberries, caught fish, hunted waterfowl and game animals, cut winter ice for refrigeration, and stored water behind milldams. Densely kettled terrains were often the principal recharge areas for the underground aquifers that kept colonial mill wheels turning at sawmills, clover mills, gristmills, grinding mills, ax mills, looms, and many other cottage industries in need of mechanical power. Damming kettles to create mill ponds sometimes didn't work very well because they leaked so much.

Kettles provided early colonials with mineral resources as well. Bog iron ore formed where the oxidized waters draining downward from sandy hills met the chemically reduced waters within and beneath acidic peat. To locate ore, colonists would do a reconnaissance for streaks of rust staining, then probe through the peat until they hit a hard clot of precipitated ore called limonite. This lumpy ore was then excavated, smelted, and hand-forged into the tools, nails, and architectural elements of early buildings and vehicles. In kettles that had been drained by geological processes, the silt liner was sought for making bricks, chinking timbers, and plastering chimneys and foundations. The purest quartz sand was used for making glass.

Cranberries—likely named for the plant's crane-shaped flower—would become the most important commercial product obtained from kettles, an industry that is still thriving.[3] Missionary John Eliot described and celebrated their benefits as early as 1647. Two centuries would pass before Henry Hall would transplant wild vines into a bog near the center of Cape Cod in North Dennis, Massachusetts, after accidentally discovering that a layer of sand blown onto a bog encouraged their growth. By 1854 the Yankee cranberry industry was thriving, with the highest production coming from the towns with the greatest density of kettles. There a landowner with soils too poor for other crops could find all of the requirements needed for a successful cranberry operation: a layer of peat beneath the basin; an abundance of coarse, clean sand; saturated conditions at depth; and a sufficiently long growing season.[4]

These conditions are common in rigorous northern landscapes like northeastern Maine, the Upper Peninsula of Lake Michigan, and the abandoned bed of Glacial Lake Agassiz in northern Minnesota, where blanket bogs cover vast areas. But within the glaciated fringe, especially to the south, the conditions for cranberry growth are met most often in isolated kettles within sand derived from granite source rocks, and in places where the peat is thick. Cape Cod had just the right sand because it lay at the terminus of the meltwater pathway draining New Hampshire, the Granite State. It had peat thick enough and acidic enough because the growth of bog plants had been able to keep pace with the steady rise of the water table being lifted by the rise of the sea. It had just the right climate as well, with sunny summers, a long growing season, and cool, foggy winters.

Cranberries would become an economic salvation for residents of southeastern New England after the Civil War. The whaling industry and the cod fisheries were in decline, clipper ships and schooners were being replaced with steam-powered vessels, and railroads had bypassed the Cape. Then, almost overnight, cranberries became economic "red gold," inspiring the eastern counterpart to California's gold fever. The demand was high in part because, being rich in vitamin C, they did for American sailors what limes did for the British—kept them free from scurvy. Cranberries were also sold as a natural tonic for good health and as an elixir for a long life.

To develop a bog, farmers would dig a perimeter ditch and another across the center to regulate the water level. Once drained, the surface was easily stripped of bog shrubs, covered by coarse sand, and planted with vines. The annual handpicking employed thousands of people, especially immigrants from Finland, whose glaciated homeland resembled southeast Massachusetts. Today cranberry production has expanded to vaster sand plains in New Jersey, Wisconsin, and Maine because most harvesting is done by machine.

OF ALL THE INDUSTRIES associated with kettle lakes and ponds, the most interesting to me was the harvesting of ice during late winter. In an age before electricity, ice blocks were the main form of summer refrigeration. Farmers had long been cutting their local ponds for ice. But when railroads appeared during the first half of the nineteenth century, that same ice could be commercially mined and exported. The Walden Pond on which Thoreau lived his "life in the woods" was also a seasonal quarry of industrial scale where Yankee capitalists called "ice kings" presided over hundreds of Irish workers living in huts along the shore. During the winter of 1846–1847, Thoreau estimated that more than ten thousand tons of this "azure tinted marble" had been cut by hand from the pond, smothered in "coarse meadow hay," loaded onto the Iron Horse, and shipped the world over. "The sweltering inhabitants of Charleston and New Orleans, of Madras and Bombay and Calcutta, drink at my well" is how he described this important export business.[5]

Any body of water that freezes thickly can be used for ice to keep storage rooms cool or bottles cold. But if the ice is to be used for chilling drinks, or to be in direct contact with food, it must be exceptionally pure. Kettle ice met this standard better than that from basins fed by runoff because its water was filtered naturally four times over. Rain and snowmelt dripped through the pines and sterilizing mulch of needles, downward through a filter of sand to the water table, sideways through aquifers, then finally through the wave-washed sands of the shore above the silt liner. Turbidity in isolated kettles was negligible because there were no inlet streams to bring in floodwaters and few muddy banks to erode. The surface water was freer of microbial pathogens because kettles were typically removed from roads and village populations where body waste fouled the way-sides. The concentration of phytoplankton was low, owing to the limited nutrient condition of the filtered groundwater. Kettle hollows trapped pockets of cold air and stilled the wind, leading to greater thicknesses of pure ice with fewer bubbles. Thoreau understood why Walden was being quarried, recognizing that its crystalline ice dif-fered greatly from the "white ice of the river, or the merely greenish ice of some ponds." During his era, a toast of ruby red cranberry juice on the rocks would have been a double celebration of the best com-mercial products that kettles had to offer.

OF ALL THE GOOD THINGS that came from New England kettles, however, the most critical was the heightened appreciation of nature that launched transcendentalism as an intellectual movement. In 1832, the same year that Henry Rowe Schoolcraft was canoeing toward Itasca, the Reverend Ralph Waldo Emerson was having the most character-defining year of his life. Beset with grief over the death of his nineteen-year-old wife, Ellen, and standing on the precipice of a personal spiritual crisis, he resigned his Unitarian pulpit and left for his first European tour. The culmination of his trip—an epiphany in the Jardin des Plantes in Paris on July 13, 1833—helped inspire his 1836 manifesto, *Nature*.[6] This small book marked a turning point for American literature, claiming that nature was, or should be, an inte-gral part of spirituality and religion. Emerson shouted these ideas during his divinity school address at Harvard's 1838 commencement, a speech that would be looked back upon as "America's declaration of literary independence."[7] Within a few years Bronson Alcott, Louisa May Alcott, Orestes Brownson, William Henry Channing, Margaret

Fuller, Nathaniel Hawthorne, Thomas Higginson, Theodore Parker, Elizabeth Peabody, Frank B. Sanborn, Henry David Thoreau, Jones Very, and many others had coalesced into a vibrant literary movement.

The cultural shift away from the rigid Puritanism of the seventeenth century toward the more flexible Unitarianism and Deism of our eighteenth-century founding fathers was a European import. Benjamin Franklin, Thomas Jefferson, and John Adams were intellectual as well as political diplomats, traveling back and forth between Europe and America during a free-trade period of ideas known as the Enlightenment. But the second step in America's spiritual emancipation—its attachment to nature—was planted directly over the kettle lake terrain of Concord, Lincoln, and Sudbury, Massachusetts. During the 1830s and 1840s, these large "ponds" were clustered within bumpy patches of pine woodland unfit for most agricultural purposes and removed from the streams that powered village mills.[8]

Geographically, to call this collection of glacial sinkholes a lake district is a bit of a stretch. But spiritually, there is no more famous cluster in America. Each basin was filled with what was holy water to the transcendentalists. Margaret Fuller had her religious conversion on the shore of a pond. Nathaniel Hawthorne wished to be baptized in one. Thoreau became psychologically "high" while floating upon Walden's tranquil surface. Together, they quaffed kettle waters straight from rippled pond surfaces. The drinking of water from kettles by intellectuals continues today because the water supply for two of the greatest universities in the region—Harvard and MIT—comes from one appropriately called Fresh Pond.[9]

"A lake is the landscape's most beautiful and expressive feature. It is earth's eye; looking into which the beholder measures the depth of his own nature."[10] Thoreau would not have written this about one of the Great Lakes, which are too vast and overwhelming. Nor would he have written this about elongate, rock-carved lakes like the Finger Lakes of New York or those in the English Lake District. Being ribbon shaped and radial in their group pattern, such lakes draw the viewer's attention outward and away, rather than inward to a single focus. Less elongate bedrock lakes of any size are too jagged in their shape, and too hard-edged to be compared with the human eye.

Several of the core ideas of transcendentalism are linked to the small, safe, soft, isolated, and accessible kettles of the Concord cluster. The quirky individuality of each—Sandy, White, Walden, Farrar, Little, Crosby, Goose, Beaver, Warners, Fairhaven—suggests a sort of self-reliance, especially in terms of hydrology. The optical and acoustic properties of Walden, being set "low in the woods," are responsible for Thoreau's utter fascination with the multisensory messages—tactile, auditory, and visual—emanating from the pond surface. He used these sensations to leverage worldly human experience to the ethereal plane, writing that the water, when "full of light and reflections, becomes a lower heaven itself so much the more important." When quiet, the water surface is "gossamer" and a "perfect forest mirror." When a breeze is blowing, it "betrays the spirit that is in the air," raising consciousness skyward.

Thoreau summarized in a poem the connection between the pond's emanations and the transcendent spirit, apparently because prose was insufficient.

*Otter Pond, Maine.*

I cannot come nearer to God and Heaven
Than I live to Walden even.
I am its stony shore,
And the breeze that passes o'er;
In the hollow of my hand
Are its water and its sand,
And its deepest resort
Lies high in my thought.[11]

The intimate physical and thermal linkages between land, water, and air within classic kettles create a special kind of beauty that no other type of lake can match. Topographically, the banks of kettles are often unusually high and steep, and are gracefully curved inward, creating a perfect stage for aqueous theater, on which "some sort of sylvan spectacle" can play upon its "liquid and trembling" floor.[12] Thoreau devotes an entire chapter of *Walden* to sounds, in part because when a layer of cool air lies above still water, sound waves travel faster and are less attenuated: amplifying slight noises, clarifying different pitches, and exaggerating echoes. Meanwhile, the shape of this theater-in-the-round fills the "surrounding woods with a circling and dialating sound . . . from every wooded vale and hillside." In the winter the mechanical stress and strain of the ice creates an endless series of restless groans, booms, snaps, creaks, and crackles. Spring temperature inversions cause a layer of warm air to become sandwiched between the cold water surface and the overlying cool air. This can create faint mists at the points of contact that refract light to an almost electric state, while simultaneously muting and muffling sounds. During autumn, the slightest change in air pressure dimples the water as if it were liquid mercury, and a single struggling insect can keep the water alive.

In a sense, kettle lakes even played a tiny role in helping America stay glued together before the Civil War. According to the literary historian Professor Alan Trachtenberg, Longfellow's *Song of Hiawatha* "was an early shot in the campaign (which included the Civil War) to imagine a continental nation with origins in the west and among the Indians."[13] This transcendental-era creation myth, based on the lake-forest Ojibwe of the northern heartland, quickly became a runaway bestseller—making its author the first financially successful poet in America and allowing him to resign his professorship at Harvard. By homogenizing the Indians, the poem helped unify the cultural polyglot of European origins that was then mid-nineteenth-century America. Henry Rowe Schoolcraft went further, claiming that native mythology provided an alternative foundation for American "literary independence," one that did not depend on Greek, Latin, or European precedents.[14]

*WALDEN* AND *HIAWATHA* COINCIDED with the rise of the glacial theory. In 1840 Louis Agassiz, a young natural scientist devoted mostly to the zoology of fossil fish, published *Études sur les glaciers*. One year earlier, Timothy Conrad had deduced that scratched rock in western New York was the work of an enormous mass of terrestrial ice that had invaded from Canada. One year later Rev. Edward Hitchcock proposed something similar for Massachusetts. Agassiz was not the first to demonstrate with compelling evidence that glaciers were once vastly larger, but he was the first to produce a full-length treatment, a two-volume monograph, summarizing an eight-year study in Europe. Building upon the work of Ignace Venetz and Jean de Charpentier, Agassiz concluded that the Alps had indeed been buried by a mountain ice cap. Soon he extended his ideas to Scotland, England, Wales, and Ireland, which he concluded also must have been covered by a coalesced mass behaving like that of Greenland.

Agassiz's book caught the cusp of change between the biblical worldview and the newer scientific one. The concept of a global freeze accompanied by death and destruction matched the catastrophic thinking of the theologians of the day, in this case the world ending with ice instead of fire.[15] Within four years, the intelligentsia was sensationalized by the 1844 appearance of *Vestiges of the Natural History of Creation*, published anonymously by Scotsman Robert Chambers due to its scandalous content. The book argued that science does a much better job explaining the origin of the physical world than does the Old Testament of the Bible, now viewed primarily as an anthology of heroic literature from the late Bronze Age Middle East. Some of the most pivotal evidence for the scientific worldview concerned the origin of water-washed glacial sediments so typical of kettle terrain, material Thoreau and his contemporaries called "diluvium" for its presumed genesis during the universal flood.

The origin of water-deposited sediment at high elevations and greatly removed from large water bodies was apparent to scientists long before they understood the glacial connection. Like the sand on modern bars, tide flats, and beaches, it was unambiguously stratified into layers, channeled by currents, and marked by diagnostic features like ripples, herringbone rills, and cross-beds. What did not make sense to early scientists was the occurrence of such features on open hillsides, often in places where gullies cut through ice stagnation terrain. Before *Études sur les glaciers*, the only explanation that made sense for such elevated water-laid deposits in mainland Europe, Scandinavia, the British Isles, and northern North America was the biblical deluge. But with the glacial theory, diluvium could be easily explained without resorting to scripture.

The catastrophic flood theory that worked so well for water-laid sand, silt, and gravel worked poorly for the enormous glacial erratics so prominent on the shorelines of kettles. The flood theory was even less adequate at explaining glacially scratched rock outcrops and the colossal boulders within till. So a compromise position was reached between scientists and theologians. The fine-grained components of till—clay, silt, and sand—had indeed been deposited during the recession of Noah's floodwaters, but the boulders and rock slabs had "drifted" into position on icebergs floating away from the poles. It is for this reason that glacial deposits are still erroneously referred to as "drift."[16]

In 1843 the American Association of Geologists and Naturalists, as it was then called, in one of its first official acts, voted down Agassiz's glacial theory, instead reaching a consensus that icebergs and a great flood were responsible for midlatitude diluvial features.[17] Incredulous, Agassiz, who had been invited to give zoological lectures at Harvard University, came to America in 1846 to see for himself. After examining the evidence throughout New England and the Great Lakes, he declared that the northern states had indeed been inundated by a single great glacier of continental proportions. Before he was to return, he accepted an offer of a professorship at Harvard, remaining there for the rest of his career and making New England the center for ice-age studies for a half-century.

Thoreau worked for Agassiz when he lived at Walden from 1845 to 1847, collecting natural history specimens for the new Museum of Comparative Zoology being founded in Cambridge. With such a close and direct link to the person most single-handedly responsible for glacial theory, it's surprising that Thoreau paid so little attention to it in *Walden*. It would be nearly half a century before the origin of Walden Pond as a glacial kettle hole became public knowledge. In 1892, geologist Warren Upham lectured on local geology to the Boston Society of Natural History.[18] One year later, naturalist John Muir made a personal pilgrimage to Thoreau's house site, writing: "Walden is a Moraine pond . . . which dates back to the close of the last glacial period when the general New England ice sheet was receding." This was confirmed in 1905 by James W. Goldthwait in his article "The Sand Plains of Glacial Lake Sudbury," published by the museum Agassiz had founded.[19] Walden may not be the most famous lake in the world, but it is certainly the most famous kettle.

THE ORIGINAL SETTLEMENT of New England had been largely about charters, loyalist patronage, and the sale of large grants of land for real estate speculation. By the time the Erie Canal opened in 1825, this eastern landscape had mostly filled up. The solution for New Englanders was to leapfrog west, generation after generation, migrating first to upstate New York south of Lake Ontario, then west to

Pennsylvania, Ohio, and Indiana south of Lakes Erie and Michigan, and then farther west to the territories generally south of Lake Superior. Although none of these places were settled as "colonies," early residents were deeply conscious of their New England legacy.

As early as 1778 a Massachusetts resident named Jonathan Carver explored the old Northwest Territory, publishing *Travels of Jonathan Carver Through the Interior Parts of North America in the Years 1766, 1767, and 1768*. His was the first English-language book to describe what would become the kettle lake heartland.[20] A few years later, a Connecticut resident named Peter Pond—who had enlisted in the French and Indian War as a sixteen-year-old—traveled up the Mississippi into Minnesota, where he spent two years trading with the Indians. His 1785 map became a significant contribution to historic geography, especially for the New Englanders who would follow.

By the early nineteenth century, New Englanders had developed a clear preference for salubrious northern climates and pine forests, and had developed considerable expertise in financing, milling, dairying, logging, and small industry. As they moved west along the northern tier of the glaciated fringe, state after state was imprinted by Yankee ways, especially Minnesota. A New England Society of the Northwest was organized there in 1856, two years before statehood. When Thoreau visited the state in 1861, he remarked that half the men he met there were from Massachusetts, and the lumbermen were from Maine. Novelist Sinclair Lewis called Minnesotans "double Puritan—prairie Puritan on top of New England Puritan; bluff frontiersman on the surface, but in its heart it still has the ideal of Plymouth Rock in a sleet storm."[21] Minnesota" remained a "New England West" until the close of the nineteenth century, electing a steady series of governors from prominent eastern families. It would not be until 1895 that the sons and daughters of European homesteaders became the dominant political force. In that year a Swede named Knute Nelson became the state's first of many Scandinavian governors.

One of the New Englanders who leapfrogged west along the glaciated fringe during the early nineteenth century was Charles Whittlesey from Southington, Connecticut. After graduating from the U.S. Military Academy at West Point and serving as a soldier during the Black Hawk War of 1832, he settled in Cleveland, where he worked as an attorney and edited the *Cleveland Herald.* Tired of office life, he made an about-face career move, working eleven years with the Ohio Geological Survey.[22] In 1848 he presented his ideas in a compelling scientific article on the drift of Ohio, following it up ten years later with a monograph far ahead of its time. Within the next few years he had mapped the southern limit of the ice sheet across the entire Great Lakes sector.

Whittlesey can be considered America's first glacial geologist.[23] His pioneering insights are numerous. He recognized that the basins of the Great Lakes were carved out of rock by glacial erosion, reported evidence for multiple ice invasions based on forest layers buried by tills, created a classification of glacial deposits based on types of material, and understood that global sea level had been

*Charles Whittlesey in the middle nineteenth century.*

lowered by the buildup of ice. Most important, he was the first to correctly interpret the origin of kettle lakes as collapse features associated with meltdown of stagnant glacier ice.[24] He recognized that there was simply no alternative mechanism that could account for all the observations.

Thomas Crowder Chamberlain, better known as T. C., was Whittlesey's successor. Born in 1843 as the son of a minister, T. C. worked tirelessly during the last third of the nineteenth century to extend mapping of the drift border and to investigate the details of its associated features. He legitimized glacial geology as an academic subject on par with the study of rocks and fossils, became president of the University of Wisconsin, then moved to the University of Chicago, where he created its first geology department. Though the concept of kettles belonged to his predecessor, it was Chamberlain who gave these distinctive landforms their permanent scientific name, borrowing it from local farmers who referred to them as "potash kettles," "pot holes," and "pots and kettles" after their shape.[25]

In 1878 T. C. published a map showing a single kettle moraine extending from near Fargo, North Dakota, to Cape Cod, a distance of nearly two thousand miles. It followed twelve large moraine loops, each created by a separate lobe of the ice sheet within the Great Lakes basins and major bays.[26] This kettle moraine did not mark the outermost limit of glaciation. Nor is it much of a moraine in most places. Instead, it is a diffuse, slightly elevated band of sandy soils, bumpy topography, irregularly shaped streams, and lake-speckled terraces deposited where the ice margin oscillated back and forth between about twenty thousand and twelve thousand years ago. It's the heart of the glaciated fringe.

THE SECOND WAVE OF IMMIGRATION to the kettle lake heartland came not indirectly from Yankee New England but directly from northern Europe. After the Civil War millions of hungry, hopeful immigrants arrived from the lowland parts of the Scandinavian peninsula and from the northern mainland on the opposite shore of the Baltic Sea. Their countries had also been covered by a large ice sheet that had left its own version of the glaciated fringe.[27] To the south lay a broad kettle moraine extending from Denmark across central Germany, northern Poland, and the Baltic countries of Lithuania and Estonia. To the north, in Norway, Sweden, and Finland, was a mix of kettles, rock-scoured lakes, and partial kettles similar to those in the Northeastern sector. Denmark, which juts west, then north from the European plain, is a glacial mirror image of Cape Cod, which juts east, then north from the mainland.

The main incentive to immigrate was the Homestead Act, signed into law by Abraham Lincoln in 1862. It gave away 160 acres of arable land to anyone willing to pay the eighteen-dollar filing fee, build a house, and farm for several years. The railroads also encouraged settlement because they stood to make a fortune shipping agricultural and forestry commodities back east, where most of the population lived. Kettle lakes were a third incentive, especially for immigrants from northern Europe.

As early as 1849 the Swedish novelist Fredrika Bremer prophesized: "What a glorious new Scandinavia might not Minnesota become!" For the Swedes there would be "clear, romantic lakes: the Norwegians would find rapid rivers; and the Danes could claim friendly pasturage for their flocks and herds."[28] Pamphlets published in many languages exaggerated the claim about Minnesota that "the whole surface of the State is literally be-gemmed with innumerable lakes . . . Their picturesque beauty and loveliness, with their pebbly bottoms, transparent water, wooded shores and sylvan associations, must be seen to be fully appreciated."[29] Norwegian-born journalist Paul Hjelm-Hansen appealed to the adventurous: "I have made a journey, a real American pioneer trip, into the wilderness, with oxen and a farm wagon . . . with a buffalo hide as a mattress, a hundred pound flour sack as a pillow, and like Fritchof's Vikings, the blue sky as a tent."[30]

The ancestors of Laura Ingalls Wilder, Ole Rølvaag, Carl Sandburg, and other literary figures heeded the call, immigrating to the kettle lake heartland. My own ancestors—Thorson, Anderson, Erickson, and Rudne—did the same in the 1870s, eventually homesteading near the edge of the prairie in southern Minnesota and well out onto the prairie in central North Dakota. My grandparents, two Norwegians and two Swedes, were the last in their line to speak Old World languages. Peak immigration from Scandinavia to the United States occurred in 1882 with 105,326 new residents.

The old New England pattern had been to hug the coast for a few generations before exploding into the interior glaciated landscape. The late-nineteenth-century immigrant pattern was to settle near the prairie-forest boundary, where grain grew well and where wood and water were easily accessed. Only in the 1870s and 1880s, and with the "advent of railroads, liberal land laws, and an enticing" market for hard spring wheat grown on the bed of Glacial Lake Agassiz and on clay-rich till, did farmers brave the flat, treeless prairies to the west.[31] The sequence is reflected in the fact that Laura Ingalls Wilder's *Little House in the Big Woods* of Wisconsin predates her *Little House on the Prairie* of Kansas.

Small lakes were critical to prairie settlement, even on the treeless moraines of the Dakotas. This is captured in Ole Rølvaag's Norwegian immigrant saga *Giants of the Earth*, the midwestern equivalent of Herman Melville's *Moby-Dick*. This tragedy narrates the physically exhausting life of Per Hansa, who moved west from southern Minnesota to the glaciated prairie in South Dakota. There prairie potholes teeming with waterfowl became important focal points in what was otherwise a lonely world where grass billowed in waves like the sea. Per was an eternal optimist, in spite of bitter winters, plagues of locusts, a wife going insane from loneliness, and perceived Indian threats. Stoic to the end, he froze to death during a blizzard while skiing for help. Rølvaag's tales of harsh winters and Norwegian bachelor farmers provided the grist for the "Wobegonesque" reminiscences of the twentieth century.

The major moraine belts of Michigan, Wisconsin, and Minnesota were generally too bumpy, too piney, and too sandy for productive grain agriculture. There the main attractions were grassy pastures and hay fields for flocks of dairy cattle. There the climate was perfect for making milk: not too hot, not too cold, but just right.[32] There milk, cream, butter, and cheese anchored a rural economy broadly similar to that of the New England interior, and of Old World Scandinavia. Wisconsin would become known as the Dairy State. Sauk Center, Minnesota, claimed to be the butter capital of the world.

By 1880 the Scandinavians outnumbered all other foreign-born residents in Minnesota, including the New Englanders.[33] There were 62,521 Norwegians, 39,176 Swedes, and 6,071 Danes. Most of the 66,592 Germans came from lands that had previously been Danish or from "White" Russia near Finland. There were 8,495 foreign-born English, 2,964 Scots, 1,103 Welsh, 7,759 Bohemians, 2,828 Swiss, 2,218 Poles, 2,272 Russians, and 1,321 French. Sadly, only 2,300 Native Americans remained. The African American population numbered 1,564. These numerical proportions were broadly similar to the other northern heartland states, though in other places there were fewer Scandinavians and more Germans, especially in Wisconsin.

Among the African Americans in the heartland was George Bonga, a legendary tracker, trapper, and fur trader in Minnesota. He was born of a black father and an Ojibwe mother and descended from Jean Bonga, who settled in the north woods in 1782, probably the first African American to do so. In Wisconsin Zachariah and Mary Morgan moved up to Pine Lake in 1870 to establish a home at a time "when no one but vanished Indians and elusive land-lookers had ever seen that part of the country."[34] Zachariah Morgan was "born in 1840, the son of manumitted slaves in North Carolina" but went to Canada before resettling in Wisconsin after emancipation. His tale of pioneering farming that involved hard work, a cabin on the lake, and a long, successful life is typical. When he died in 1894, he was both a leading citizen and a prosperous farmer.

· · ·

TIMBERING HAS ALWAYS PLAYED a role in the New England economy, beginning with the selective cutting of the tallest pines for Her
Majesty's ships' masts. The oldest sawmill for which there is a certain
date (1631) is in Berwick, Maine.[35] Timber for ships, colonial villages,
and early American houses typically came from easily accessible
nearby forests, especially virgin stands of white pine. But industrial-
scale logging for export was generally a nineteenth-century phenomenon, initially restricted to the hard-rock, north-woods terrain of
Maine and New Hampshire, where large river systems, bedrock
rapids, and protected seaside ports made it commercially feasible.
After cutting, logs were transported by river via rafts, milled on rivers
at rapids, and shipped from tidal estuaries. Most of this activity was
unrelated to kettle lakes because it took place on ice-scoured bedrock
terrains. In the heartland states, however, timbering took place almost entirely on forested kettle moraines. In this band, majestic
stands of white and red pine grew in sunny dry soils near the shores
of countless lakes generally devoid of the human presence.

Ernest Hemingway, in his characteristic spare style, described a
good stand in northern Michigan: "There was no underbrush in the
island of pine trees. The trunks of the trees went straight up or
slanted toward each other. The trunks were straight and brown without branches. The branches were high above."[36] Such pines became
the logs that were milled into boards that became the houses that
would shelter the booming U.S. population during its industrial surge
after the Civil War. Rebuilding Chicago after its great fire of 1871 was
an early stimulant to the industry.

Timbering followed the model used by the fur trade: Find it. Take
it. Export it. Don't worry about the consequences. As with the fur
trade, the Indian way of life was hard hit, even though timber provided temporary employment. Native author Winona LaDuke wrote
of her White Earth Reservation in northwestern Minnesota:

> Our creation stories, culture, and way of life are entirely based on the
> forest, source of our medicinal plants and food, forest animals, and

birch-bark baskets . . . In 1874, Anishinaabe leader Wabunoquod said, "I cried and prayed that our trees would not be taken from us."[37]

This didn't stop Weyerhauser and other major lumber companies from clear-cutting nearly every tree on the reservation at the turn of the century.

Many entrepreneurs started timbering businesses. The Minnesota state archive, for example, has records for "some twenty thousand different log brands," each marking the logs owned by a different individual or group.[38] The goal for every group was to locate the best stands, secure the timber rights, mobilize for the clear-cut, and then move on. The best logs were mature, even-age stands of pine, usually on the slightly elevated moraines. The ideal method was to work uphill from the side of a lake, timbering the slopes, rolling the logs down to the water, and floating them to a spot where they could be loaded onto gigantic horse- and ox-drawn sleds or, better yet, to a rail spur. Large kettles—especially those connected by sluggish streams—were strongly preferred because access was easy on the frozen lake surface and because the logs could either be skidded out in winter or floated out in the summer. The broad fringing wetlands of kettles that had been so important to native subsistence during the Archaic and Woodland periods became a barrier to logging operations, being too wet to skid logs over but not wet enough to float them along. Hence, if a lake had an outlet, it was often dammed to raise the water level just enough to transport logs effectively. Lakes too far off the beaten track of skid road, river, and rail were never touched.

Logging leapfrogged from east to west—Maine, New Hampshire, Ontario, Michigan, Wisconsin, and Minnesota—as the commodity was exploited. In each spot, the industry rose slowly, accelerated as competitors found a piece of the action, reached a plateau when the cut from new arrivals was offset by the difficulty of finding more timber, and then crashed as financiers jumped to the West. The timber industry migrated in phase with the local demand for frame houses in rapidly growing cities: Buffalo, Columbus, Toledo, Detroit,

Chicago, Minneapolis–Saint Paul, and Fargo. The final jump was over the mountains to the Pacific Coast, where tall, even-age stands of old-growth Douglas fir, Sitka spruce, and California redwood could be more efficiently cut, rolled down to the nearest river or fjord, and floated to the mills.

Nationally, the timber boom peaked in 1907 with 89,200,000,000 board feet.[39] Minnesota, the most western timber state in the fringe and the latest to be developed, had already peaked by then, with close to two billion board feet produced in 1899.[40] This amount, which had come from a single state in one year, had required nearly one hundred thousand railroad freight cars to transport.

Most of the timber cutting took place during the winter, when there were fewer bugs, skid trails were solid with ice, logs could be gathered at lakeside before the spring freshet, and sweat didn't smell so bad. Loggers lived either in bunkhouses or in shanties not much bigger than wigwams, and with a similar roof hole to let out smoke. Beyond the ubiquitous ax and saw was the "peavey," a pointed, hooked tool used to move logs around. The "go-devil" was a heavy, log-hauling sled with two sections separated by a pivoting axle. The "chopper" was the man with the ax. The "timber wolf" was the cruiser who marked trees for cutting. Cooks were "hashslingers," clergymen "sky pilots," the foreman a "kingpin." Meals were served in the "chuck-house." Tea was "swampwater." Getting drunk was "kegging up." Sweat was "Swedish steam."[41]

BRAVERY, WILDERNESS SAVVY, and endurance were the hallmark traits associated with the earlier voyageurs and explorers. Pioneering agriculture evoked duller, more family-based images that didn't lend themselves well to tall tales. The mining industry was associated with machinery of great power, strength, and capacity: draglines, excavators, trucks, cargo ships, trains, conveyors. For these three activities, mythical superheroes were unnecessary; reality was sufficient. Timbering the big pines was different: With no reason to celebrate courage,

independence, ingenuity, and military skill, the only thing to be celebrated was brute labor in the woods at a scale now difficult to con-
ceive. Paul Bunyan would symbolize that Herculean labor.

Based on extensive oral histories taken from those who actually
worked in turn-of-the-century timber camps, the historian Agnes
Larson concluded that Paul Bunyan was never part of folk mythology. Instead he was pure fiction, springing out of whole cloth from
the imagination of James McGillivray, a journalist for the *Detroit
News Tribune*. The debut story featuring Paul's larger-than-life
characteristics was published on July 24, 1910, three years after the
industry began its decline in the northern heartland. From this
simple beginning grew one of America's most enduring folk heroes,
now claimed by every top-tier state from Maine to Minnesota.
(My hometown of Bemidji, Minnesota, operates a twenty-four-
hour live Internet feed of a giant statue of Paul and Babe erected in
1937.) Ironically, Paul's animal companion, Babe the Blue Ox, was
probably more authentic than he was.[42] Everything about Babe's
story—the gargantuism, the geographic setting, and especially the
blue color—is consistent with a fossil discovery from a northern
kettle bog.

Recall that ice-age life, especially for herbivores, was oversized.
Mammoths stood taller than the largest African bull elephants.
Beavers were the size of hogs. Pleistocene bison were bigger than
their modern counterparts, their horns extending nearly six feet
across, not counting their sharp protein tips, which have long since
decomposed. One such extinct bison was found at Nicollet Bog in
Itasca State Park, Minnesota, in 1937 when men digging the foot-
ings for a new road bridge encountered iron-stained bones near the
base of the peat.[43] They were from *Bison occidentalis*, a long-horned
variety that would later interbreed with *Bison antiquus* to give rise
to the smaller, short-horned bison we know as *Bison bison*, the buffalo. At this site, Archaic hunter-foragers apparently drove several
bison into the boggy reach between two kettles until they became
mired and were easily killed. After they were butchered in place, sec-

*Paul and Babe in Bemidji, Minnesota.*

tions of the meat were hauled to the top of the bluff, roasted, and feasted upon.

Although the ox from Nicollet Bog was not blue, it could have been. A true-to-life blue Babe was found by miners in Alaska who were removing peaty mud to get at the underlying gold-bearing gravel. (See image on p. 48.) The chalky blue color of its carcass was due to a mineral named vivianite, an iron phosphate produced under fairly rare chemical conditions.[44] The sediment must be phosphorus poor, making the skin and bone the dominant source. The water must be iron rich, facilitating the reaction at the contact between peat and skin. The oxygen content, pH, and temperatures must all be low, in order to enhance crystallization. I have seen this vivianite-stained Babe in the flesh. I have also seen this blue mineral as chalky coatings on fossils elsewhere in Alaska, faint layers at the base of New England bogs, and grains in the base of kettle lake peat. It has also been widely reported in heartland lake sediments: for example, in Elk Lake, Minnesota, close to Itasca.

A fossil origin for the blue Babe legend is consistent with the typically serendipitous discoveries of strange things in the rural country. Excavation and construction bring "working men" into contact with the evidence of Pleistocene life in remote locations. Typically, those who discover the bones have little scientific education, and are more likely to invent a creative explanation for something unusual—like an enormous blue ox in a bog—than to ask a museum curator. This was especially true during the early twentieth century, when graduating from high school was a significant accomplishment many families could not afford.

LUMBERING HAD IMPORTANT ECOLOGICAL impacts on kettle lakes.[45] Some filled with mineral sediment delivered by spring runoff from deeply frozen skid trails and logging roads. Logging slash decomposed during summer, sending a steady trickle of nutrients into the water for years. Clear-cutting removed the shade, drying up the soil, increasing the chance for lightning and accidental fires. One such fire in Hinkley, Minnesota, near the peak of the logging boom killed more than 418 people in 1894 when more than three hundred thousand acres burned. Those who survived sought refuge in a small kettle called Skunk Lake. Something similar happened in Wellfleet, Massachusetts. Wildfires such as the one in Hinkley created nutrient-laden ash, which quickly leached into lakes, causing algal blooms. Lakes raised by dams to get logs submerged adjacent marshes, removing a sink for nutrient. As a result, most kettle lakes became polluted during the logging period, growing murkier and weedier than they would otherwise have been. Meanwhile, lakes untouched by loggers remained as pure as ever.

EVERY GLACIATED STATE has its own rock quarries. New England and the Superior Province are renowned for their granite and their slate. New York, Pennsylvania, Ohio, Indiana, Illinois, and Iowa are renowned for their limestone and sandstone, which are cut into dimension stone for building and flagstone for paving. These can be big businesses, though nothing on the scale of the iron- and coal-mining industries, which fueled America's rise as a superpower.

Gilded Age capitalists living in the corridor between Pittsburgh and Boston were interested in a commodity that was rare enough to be monopolized, and which required huge infrastructure investments. Other than coal and oil, their most notable target was iron ore from the Archean rocks of the Superior Upland. Minnesota's earlier grain-milling industry had been centered in Minneapolis and locally owned and operated. In contrast, the Minnesota iron business was run out of eastern cities by the likes of Andrew Carnegie, J. D. Rockefeller, Alfred Munson, and Charlemagne Tower. They had the legal and banking savvy to acquire the mines, the capital to construct the port facilities, and the ships to bring ore and coal together in places like Cleveland, Ohio; Gary, Indiana; and Pittsburgh, Pennsylvania. The ore came through the ports of Marquette and Munising, Michigan; Ashland and Superior, Wisconsin; and the greatest of all, Duluth, Minnesota. Each major source of ore was referred to as an iron range, though none were high enough to be considered proper mountains.[46]

Mining began rising in importance shortly after the Civil War. A small gold rush in 1865–1866 in Vermilion, Minnesota, was likely a hoax but helped shift a focus to mining.[47] Iron ore was discovered in Ironwood, Michigan, in 1871; the railroad came in 1884; and by the turn of the century the place was flooded with immigrants from overseas looking for work, many from Finland. In Ironton, Michigan—not to be confused with Ironwood—the Pine Lake Iron Company of Chicago opened a pig iron plant in 1881 that closed within a few years. Such local boom-bust stories were repeated over and over across the southern shore of Lake Superior.[48]

The pivotal event for the iron industry was the discovery of its mother lode, the Mesabi Range in the mosquito-infested and thickly wooded forests between Duluth and Grand Portage. There three Merritt brothers—Leonidas, Casiaus, and Napoleon—"stumbled on what would turn out to be the single most valuable mineral deposit in the history of this continent. In the swampy uplands near the headwaters of the St. Louis River, just a few feet under their boots, lay a lode of iron ore rich beyond imagining; in places the soft red hematite rocks were actually strewn about in plain sight."[49] Open-pit mining began in 1890. Peak Minnesota production of 2,484,854,372 tons came in 1960. By the time it was over, the "Merritts lost their mines and their railroad, dying as poor men, their red ore being fed into the furnaces of millionaires."[50] Within the last half of the twentieth century, the industry was forced to switch to a lower-grade ore called taconite. Today the iron range is nearly a bust, the victim of new materials, technology, and international competition. In its wake is a locally decimated landscape. The sons and daughters from mining towns—people like Bob Dylan, from Hibbing, Minnesota—left for better opportunities elsewhere.

THE DIFFERENT GEOGRAPHIES associated with grain agriculture, timber, and mining would play an important role in setting the stage for the next phase of history in the kettle lake heartland. Grain agriculture established and sustained human communities to the south and west along the edge of the prairie. Iron mining did the same to the north, along the southern edge of the Canadian Shield. Between these economic bands the land was dominated by lakes and kettle moraines and populated by scattered dairy farms and dozens of towns with dwindled populations left over from the logging boom. Places such as Grand Rapids, Minnesota; Rhinelander, Wisconsin; and Cadillac, Michigan, had boomed up from trading posts and road crossings before busting back down when the logging camps closed. One example is Seney, Michigan, the setting of Ernest Hemingway's great fishing story "Big Two-hearted River"; it had thirteen saloons during the peak of the boom. Within a few years, Seney had burned to the ground, leaving only a few residents remaining.[51] I drive through it on my annual trip between New England and Minnesota.

In the ravaged woods between these boom-bust logging towns, forests of jack pine and birch were regrowing. Accessible forests in lakeside settings such as Itasca were being turned into parks. During the 1930s these abandoned lands were repopulated by a second transient surge of population associated with the Civilian Conservation Corps, a make-work project for unemployed breadwinners during the Great Depression. With the forests regrowing, the lakes recovering, the mines closing, and farm populations falling, the stage was set for the use of kettles as aesthetic and recreational resources.

7.

# Family Lake Culture

THROUGHOUT THE NINETEENTH CENTURY, Americans had been shifting away from rural outdoor occupations—farming, mining, forestry, construction, and fishing—toward indoor ones in cities. One consequence of this trend was the desire to have access to both environments: to live in the bustle of the cities, with all of their opportunities, yet to experience rest and renewal in the country.[1]

The recreational enjoyment of lakes began in the early nineteenth century in the northeastern United States, especially within the New England interior, the Hudson Valley, the Adirondacks, and the Catskills. After the Civil War, middle-class wealth and expectations began to rise dramatically; less work was being done outdoors on the farm, more inside sweltering factories and offices. Given the rising demand for passive outdoor recreation, stagecoaches on turnpikes and iron horses on tracks began bringing urban tourists to particularly scenic spots. Lakeside resorts, often with grand Victorian hotels, were built on destination lakes, such as Lake Winnipesaukee, New Hampshire, and Lake George, New York. Resorts also emerged near thriving midwestern cities, provided that transportation was available. Lake Geneva, Wisconsin, for example, catered to wealthy summer residents from Chicago, claiming to be the "Newport of the West."[2] Large, accessible kettles like Lake Ronkonkoma on Long Island and Walden Pond near Boston were also developed for public tourism.

Tourist hotels were never built along the shores of Walden Pond; they weren't needed because the Boston-Fitchburg railroad could carry visitors to and from the city on day trips. Coming out from Boston for the afternoon was so easy that by April 1854, just a few months before the publication of *Walden*, Thoreau witnessed "gentlemen and ladies sitting in boats at anchor under parasols on the calm afternoon."[3] In 1866, only four years after his death, a recreational development project called Lake Walden opened for public use, complete with beach improvements, bathing houses, picnic grounds, swing sets, a boat rental facility, landscaped gardens, an engineered walkway around the pond, and later a carnival-style amusement park. Though Thoreau wrote *Walden; or, Life in the Woods* as a manifesto for self-reliance and natural spirituality, he unwittingly played a role in drawing mobs of tourists from urban Boston to the pond.

Thoreau's life at Walden was no wilderness experience like that of the early explorers and fur traders who lived removed from civilization, often in rude cabins, and fed themselves from the land. Instead, he lived in a tidy cottage finished on the outside with secondhand siding and on the inside with plaster. The house site was on privately owned lakeshore within easy walking distance of town, where he often went for family meals. Much of the nearby woods had already been logged off. Thoreau's best friend and first biographer, William Ellery Channing, went so far as to claim that Thoreau never really lived at Walden, but "bivouacked" there at night.[4] When Thoreau wasn't writing, botanizing, "sauntering," or hoeing beans, his principal outdoor activities at Walden were swimming, boating, fishing, and just hanging out in a beautiful place, which is what most folks do today. It's no surprise that Thoreau claimed to have had more company when living at lakeside than he did when living in society. This is also true of many families today, who experience maximum social comfort during summers at the lake.

More than ten years after his sojourn at Walden, Thoreau developed a cough. Chronic bronchitis turned into fatal tuberculosis. Some say he had become exhausted while impetuously chasing a fox through the woods. Others say he got too cold while distractedly counting tree rings on stumps. Unable to shake his illness, Thoreau was advised by his physician to spend some time in a different climate, perhaps the West Indies. Instead, Thoreau surprised his friends and family by going to Minnesota, where the air was alleged to be therapeutic, and where he could recuperate in a cluster of kettle lakes like those he fell in love with in Concord. After arriving at Saint Anthony Falls via steamer on May 25, 1861, he headed straight for Lake Calhoun, a "pretty little lake."[5]

Today this kettle is a crown jewel inside urban Minneapolis, equivalent to Jamaica Pond in Boston. But in 1861 it lay in a woodsy setting within easy walking distance of the city, which then had a distinctly New England flavor. From Mrs. Hamilton's boarding-house on its shore, Thoreau did what he had done at Walden: breathed the pine-scented air, watched the play of light, listened to the sounds, swam in the water, botanized in the woods and swamps, and explored the habits of streams. After two weeks of respite, Thoreau described himself as feeling "considerably better than when I left home." Nevertheless, he returned to Concord without having shaken his consumption and died in August the following year.

Recreational development of small lakes for family second homes took off in New England ahead of in the heartland, due in part to a serious recession in agriculture and the growth of urban industry during the final decades of the nineteenth century. Many ponds that had formerly been used as watering holes for farms and reservoirs for mills were being reclaimed by forest. During the summer, the sons and daughters of farmers, now working in cities, were lured back to the country by the promise of fresh air and rural tranquillity. Conveniently, a network of carriage roads from the vanishing Yankee-village era led outward from urbanizing areas. Soon privately owned lakeshore was being sold to nearby families for the purposes of seasonal rest, relaxation, and respite. Lake Boon, a dozen or so miles southwest of Concord, began to boom in the first decade of the twentieth century.

The interest in lakeside property started small and with just a few sportsmen asking the farmers who owned land around the lake if they could erect cabins for the summer. These farmers, sensing more growth, began building cabins to rent to summer visitors. As the years went by, the number of people visiting the lake increased and cabins grew to more substantial cottages.[6]

Such activities were not yet possible in the less settled, more wide-open spaces of the Midwest, where only a few kettles were reachable by road. There popular use of lakes for recreation did not occur until the Ford Model T was mass-produced in 1913, making personal transportation both convenient and affordable. No longer were city dwellers restricted to places reachable only by railroads, ferries, and stagecoaches. Instead, the concentrated power of gasoline brought vast areas within easy reach of heartland cities. The power of diesel oil fueled the bulldozers, rollers, and graders that created thousands of miles of new gravel roads, and powered the scoops and excavators that helped develop lakeshore cottages. Kerosene made lighting and heating more efficient than burning candles and chopping wood. Coal brought electrical power plants, which made lighting and heating even more efficient and refrigeration, radio, automatic water pumps, and kitchen appliances possible. A transportation and tourist-support infrastructure consisting of state highways, county roads, gasoline stations, and repair shops emerged. With a tank full of gas, families could now pack up the kids, head for the country, and spend quality time without suffering the privations of pioneering life, or losing touch with the rest of the world. They could have their pastoral cake and eat it too.

The rise of an egalitarian, middle-class lake culture associated with midwestern kettles was captured by Sinclair Lewis in the background of his 1920 blockbuster novel *Main Street*, which was based on the author's own personal experience.[7] The novel, alleged to be third in popularity after the Bible and *Ben Hur*, is so loaded with exquisite lake imagery that it sounds *Walden*esque: "The Lake was garnet and silver" and "The wrinkled water was like armor damascened and polished."[8] In 1916 Sinclair and his wife, Gracie, bought a car in order to explore the undeveloped, lake-dotted countryside still in the process of being logged off. For their sojourns, they made a heavy expedition-style tent large enough for a table, air mattresses, and all the gear they would need for a few days of adventure at the lake. Included were a bird book, a wildflower field guide, and a camera.[9]

They made trips to Lake Itasca and the clusters of lakes near Bemidji, Fergus Falls, Mankato, and Pequot Lakes, all places that T. C. Chamberlain had mapped as kettle moraines. In spite of carrying an ax and a shotgun, they were not pioneers, but part of a growing trend of back-road travelers to back-country lakes, equipped with steel cable and a shovel for getting unstuck. Also in 1916 Congress first set aside matching funds for state highway construction. Within a few years, the Ten Thousand Lakes Association of Minnesota had been created to promote what was becoming the state's third most precious resource, after its fertile agricultural prairie and the mills near Saint Anthony Falls. By 1925 there was so much interest in "motor camping" that one could subscribe to a magazine called *Motor Camper & Tourist*.[10] Scarcely a half-century earlier, parts of Minnesota's and Wisconsin's thickly wooded lake country had been deemed of so little value that they had been set aside for Indian reservations.

In general the growing streams of traffic headed northward, which is why the colloquial phrase "up to the lake" is used, rather than "out" or "down" to the lake. Residents of the Red River valley on the Dakota-Minnesota border headed generally east-northeast. Residents of Des Moines and Minneapolis went practically straight north. The exodus from Chicago was north to Wisconsin, northwest to Minnesota, or east into Indiana before curving up into Michigan. Those fleeing Milwaukee's summer heat either doglegged west to the central part of Wisconsin or northeast to Michigan's Upper Peninsula. Detroit residents headed either straight north into Michigan or east into southern Ontario, where they competed with the crowds from Canadian ports like Toronto. Even those from "Middletown" (Muncie,

Indiana), the sociological archetype for small cities during the early twentieth century, felt the pull of lakes to the north.[11]

Heading in a northerly direction usually meant crossing the boundary between the patchwork of farms and deciduous forest to the south and the more continuous coniferous forests of the north. Sterling North, an author from my birthplace of Edgerton, Wisconsin, described this ecological transition: "Now the farm odors and fragrances blended into the great perfume of the north woods: the sharp spicy aroma of the firs, and the fine hot scent of the pine needles lying four inches thick to blanket the forest floor."[12]

Part of the appeal of heading north was to reenact the primitivism of earlier centuries. Ernest Hemingway translated this sentiment into literature because he was imbued with it as a child. Carlos Baker's exhaustive biography of Papa begins with this line: "As soon as it was safe for the boy to travel, they bore him away to the northern woods." Their trip was from Oak Park, Illinois, near Chicago, to their summer cottage on Bear Lake, a scene of pine forests, sandy beaches, and rowboats against the dock.[13] The Hemingway cottage was quite small, with a stone fireplace for heat and kerosene lamps for lighting. Their toilet was an outhouse, their bathtub the lake. From this base, the adolescent Ernest went fishing, experimented with quaking bogs as slow-motion trampolines, lusted after Indian girls, and spent countless days becoming a naturalist.

In an age when wealthy easterners went on safari to Africa, middle-class folk from the Midwest went north to hunt deer and catch trophy fish. Hunting and fishing camps usually began as clusters of frame canvas tents at ends of old logging roads. Soon guide services were creating semipermanent camps where rude shacks and cabins provided overnight accommodations. They tended to spread their operations widely over the landscape, often one to a lake, because primitivism was a big part of the draw. These guiding camps quickly evolved into small resorts built with families in mind, after the addition of a place for women and children to be while the men went fishing. A small sandy beach with a dock was essential. It wasn't long before the family resort experience became an annual habit, especially for those who didn't want to be troubled with taking care of lake property or who felt too alone in the woods by themselves.

Lake culture was focused less on the magnificence of grand landscapes and a wilderness experience than on the simple act of getting away from it all, preferably with friends and family, and at a place you could call your own. Spending time at the lake cottage was less a vacation than an altered state of existence, an emotional bubble universe separate from the rest of the year. Ovar Lofgren, an anthropologist of middle-class life, wrote that "one of the most striking characteristics of these cottage cultures has to do with rhythms and temporalities" and "ritual comings and goings." Cottages are indeed "dream spaces, because most of the year they are inhabited only by longings and memories."[14]

The writer E. B. White poignantly captured the sentiment in his 1942 memoir essay "Once More to the Lake." He wrote of his boyhood experiences at a small partial kettle of the sort largely bypassed in the previous century.

> Summertime, oh, summertime, pattern of life indelible, the fadeproof lake, the woods unshatterable, the pasture with the sweetfern and the juniper forever, and ever, summer without end . . . the cottagers with their innocent and tranquil design, their tiny docks with the flagpole and the American flag floating against the white clouds in the blue sky, the little paths over the roots of the trees leading from camp to camp and the paths leading back to the outhouses and the can of lime for sprinkling, and at the souvenir counters at the store the miniature birch-bark canoes and the postcards that showed things looking a little better than they looked. This was the American family at play.[15]

This was my own family too. In 1928 my father's Swedish American parents bought a small cottage on the shore of Union Lake in northwestern Minnesota just a few miles from their flatland home in Fertile, Minnesota, located on the pool-table-flat former lakebed of

Lake Agassiz. The purchase price was five hundred dollars. Then only five years old, my father was the perfect age to have the taproot of lake culture planted into his family tree. After he grew up, married, and started a family, I was born in a small Wisconsin town located within a kettle moraine near the shore of Lake Koshkoning.[16] One of my first memories is its sheet of blue water, rimmed by white sand and speckled by cottages. Another early memory is of the red-wing blackbirds, green frogs, and dragonflies living around a prairie pothole in Manfred, North Dakota, where the Norwegian side of our family had homesteaded in 1892.

Shortly after the Great Depression, heartland cities and towns such as Youngstown, Cleveland, Akron, Toledo, Pontiac, Flint, South Bend, and Gary became what John Steinbeck called "great hives of protection." The "roads squirmed with traffic; the cities were so dense with people . . . an electric energy, a force . . . so powerful as to be stunning in its impact . . . vitality was everywhere."[17] Such energy increased the demand for rest, relaxation, and family-focused gatherings. It was in this context that kettle lakes would be most intensively used by Americans: for their mental health benefits rather than as a source of food or marketable commodities. In a nation industrializing between two world wars, lakes were to become of "great significance to psychiatry," as the pioneering limnologist G. Evelyn Hutchinson noted.[18]

During the baby boom years of the 1950s and 1960s, my family lived in one small midwestern city after another. But we spent at least part of every summer at the cottage on Union Lake. There I adapted to lakeshore life by becoming amphibious. My hair bleached to reflect the sun's rays. My skin turned bronze to prevent a burn. My toes became antennae, able to differentiate the tickle of grass, the squish of marsh mud, the graininess of beach sand, the wooden planks of the dock, and—after a joyous leap—the cool freshness of the lake.

I recall the rowboat tied up against the ramshackle dock, with its 3-horsepower Evinrude motor and gas can. Life jackets were draped on lawn chairs. Fishing poles were strewn about, their lines tangled with bobbers and fishing lures with fascinating names like Daredevil spoons, Mepp's spinners, and Hula poppers. The outer limit of child habitat was the ominous "drop-off," beyond the end of the dock where the water deepened abruptly. Men—sitting like statues in boats— would cruise along it, trolling for trophy fish called lunkers. Inland from deepwater was a set of concentric play zones, each with its own activity. The first was for swimming and floating on inner tubes and rafts. Next was the splash zone, where kids could run, fall, wrestle, and leave a mishmash of small footprints on the submerged sand. Rising up from the miniscule beach was the sweeping lawn where we tumbled and rolled and played badminton, the screen porch with its slapping spring-loaded door, a one-story cottage no bigger than many modern garages, the work shed with its fascinating ax grinder, the outhouse with its newsprint catalogs and bucket of lime, and the woods with shade and rich earthy smells. Gender-oblivious packs of preadolescent children roamed in and out of these play zones with great expectations, never quite sure what they would find. At a nearby family resort we would spend our allowances on bottles of ice-cold soda pop and pieces of either Bazooka or Dubble Bubble gum. I don't remember any cottages having televisions. Instead there were board games, jigsaw puzzles, a sagging badminton net, and, most important, no schedule for hours a day. Lake life was the best part of boyhood life.

HISTORIAN WILLIAM CRONON, in *Nature's Metropolis*, his scholarly book on the history of Chicago, opens and closes with similar personal childhood reflections from Green Lake, Wisconsin.

> Green Lake had nothing very wild about it, but it was a lovely place to be a child. We swam and canoed. We played croquet and softball . . . We rambled through the woods behind the cottage, collecting butterflies and leaves. Many nights, we walked the mile down the lakeshore to the Terrace Grocery, where we bought ice cream cones before heading home to watch the rising moon create rivers of light across the surface of the lake . . . it would be hard to imagine a luckier, happier place in which to grow up.[19]

His sentiments are representative of a pan-galaxy lake culture that flourished in the mid-twentieth century, especially in the northern heartland. Those of E. B. White are representative of my father's generation, yet are similar to those of Cronon and I, suggesting that little had changed in the first half-century.

Kettle lakes were ideal for young families. The waves were non-threatening, the shorelines were sandy, a surface layer of warm water developed fairly quickly, and there were neither sharks nor jellyfish for mothers to fret about. Larger bedrock lakes and the ocean had rougher waves, harder edges, proportionally fewer beaches, and colder water. Public parks that featured kettle lakes were among the most intensively used in the United States. For example, by the 1930s Walden Pond was attracting more visitors than now, up to fifteen thousand per day. In the heartland the tremendous number of lakes, their chaotic distributions, and their irregular shapes offered a sense of seclusion greater than that of places where there were fewer, more regular-shaped lakes.

Families who had become hooked on lakes and had the resources took the plunge and bought a strip of lakefront property. Cottages could be enlarged as families expanded and as additional resources became available, perhaps an entry porch, an indoor toilet, and a boat-house. Zoning was unthinkable. A landowner or local real estate agent would simply buy a chunk of lakefront, divide it into narrow strips resembling piano keys, clear and bulldoze a small access road, and sell the lots. Ad hoc neighborhoods emerged automatically because lake perimeters were usually segmented by exposed peninsulas, broadly curved reaches with beaches, and marshy bays. In other cases church groups, city neighborhoods, fraternal organizations, and extended families would buy a block of land or an island and build their own summer community.

Potable water was plentiful and easy to find because the sand bordering most kettles filtered the water and created a gently sloping, very predictable groundwater table. Wells were easily dug with a shovel, drilled with an auger, or hammered in by driving a well point downward into the pliant, pebbly sand. Before electricity, the water was raised by creaky hand pumps made of cast iron or hauled up from the lake in buckets. With electricity, wash water was pumped up directly from the lake, to which it was often returned; alternatively, it was drained onto the sand. Outhouses were moved as they filled up. On the typically narrow lake lots, this usually meant farther back.

Cottages on lakes near anchor cities like Minneapolis, Indianapolis, and Milwaukee gradually became swallowed up by urban and suburban expansion, becoming year-round houses surrounded by city blocks or curved suburban neighborhoods. Small towns on outlying lakes grew up to support the emerging tourist economy, many tripling in size during the summer. The proprietors of locally owned businesses—grocery, lumber, and hardware stores, repair shops, banks, taverns, and bait sellers—had the opportunity to live on lakes year-round. During the Great Depression, tens of thousands of unemployed men were sent up-country through the Civilian Conservation Corps to build the infrastructure of parks, public forests, and scenic roads, mostly out of pine logs. When this public works program ended, many chose to stay, resettling their families in the lake country.

THE MIDCENTURY MIGRATION of American families toward the woods and lakes was a partial reenactment of the American frontier experience. In the pioneering phase of lake culture, small groups of men from dissimilar experience roamed the countryside fishing, hunting, and trapping together.[20] Its settlement phase brought together people with a wider range of religious, ethnic, and economic backgrounds than was found in the city neighborhoods from which they came. Rich kids from out of town played with poor kids from town. Sons and daughters of plumbers and bankers fished together. Kids from Detroit played cards with those from Chicago. Informal learning took place through an appreciation of nature.

"Camp," "cabin," and "cottage." These are interchangeable names for the same dwellings, though the connotations and regional pattern of use varies. The word "camp" connotes a primitive structure in a rustic, woodsy setting, perhaps a frame tent or a rude shack, and without a lawn; whereas "cabin" evokes a log structure, perhaps with a sand road and a clearing out front. A cottage is likely built of frame lumber with at least some architectural trim and perhaps a flower box, and in a setting better described as pleasant than rustic. The color for a cottage probably would be lighter than for a cabin, probably a pastel to match a sunnier, more yardlike setting. "Camp" is the dominant word in northern New England, "cottage" to the south. In the heartland, "camp" is hardly ever used. There practically everything is a cabin if it's rustic, regardless of what it's made of. If it's not rustic, then it's normally a cottage. Actually the dwelling is often not mentioned at all. One just heads "up to the lake" if it's a trek and "out to the lake" if it's local.

The names given to lakes also say something about lake culture. In the upper Midwest, lake names tend to be simple and descriptive, rather than ostentatious or pretentious. I know of no "Great Ponds" in the heartland, but there are dozens in New England. In Minnesota the ten most common lake names are, in descending order: Mud, Long, Rice, Bass, Round, Horseshoe, Twin, Island, Johnson, and Spring.[21] Spring Lake hints at the groundwater regime of kettles. Round Lake and Long Lake are comments on the full spectrum of shapes, from circular trapdoor kettles to the ribbon-shaped kettles within subglacial tunnel valleys. Moderate fertility—rather than crystal clarity—is suggested by the names Mud, Rice, and Bass, while Horseshoe, Twin, and Island reveal the random topography of coalesced kettles. Johnson Lake reflects nineteenth-century Scandinavian immigration. Curiously, there are no native names in this list.

FAMILY LAKE CULTURE SPANS the entire glaciated fringe, developing wherever small lakes are isolated from the buzz of modern life and can be accessed by road. They need not be kettles. Any small lake will do, provided there is a sense of outdoor intimacy rather than grandeur, security rather than adventure, enclosure rather than endless vistas, neighbors rather than gated communities, and water warm enough to swim in. The view is usually similar to that seen from the shores of Walden Pond—a limited stretch of blue water bounded by a forested bluff, something that can be safely crossed in a few minutes by boat. Such conditions are present in all nineteen northeastern and north-central states above the ice-sheet border. But only in the northern heartland of Michigan, Wisconsin, and Minnesota are there tens of thousands of lakes within a day's drive of large cities, and in a setting where the lake aesthetic is not diluted by grander elements of the landscape. There middle-class recreational lake culture became the dominant part of the regional recreational identity.

In New England the beauty of the seascapes and mountains overwhelms what its many kettle lakes and ponds have to offer. The charming Cape Cod towns of Eastham, Truro, Wellfleet, Chatham, and Brewster, for example, contain dozens of stunningly beautiful kettles that are every bit as nice as those of the heartland. Yet the vast majority of residents and tourists ignore them, their attention drawn to the sea. In his book on Cape Cod, Thoreau paid practically no attention to the beautiful Walden-like ponds that commanded nearly his full attention in Concord.[22] Lake Chocorua, New Hampshire, provides another case in point. This stunningly beautiful kettle lake—arguably New Hampshire's finest—is admired most for the way it reflects and magnifies the Presidential Range beyond its far shore. Kettles are so downplayed in New England that the bestselling tourist guide *Water Escapes in the Northeast: Great Waterside Vacation Spots* doesn't list a single one out of the thirty-six popular destinations profiled.[23] In a similar book about the heartland with the same title, most of the profiles would probably be about kettles.

A neighborly family lake culture has not flourished around the Great Lakes because they are too much like the sea.[24] Herman Melville wrote of their "ocean-like expansiveness, with many of the ocean's noblest traits." Poet and novelist Ana Castillo, born and raised in Chicago, described Lake Michigan as "Mediterranean Lite . . . while it looks like the sea it is not the sea . . . Rub your eyes and look again, it's a lake."[25] The five Great Lakes hold twenty times the freshwater of all other U.S. lakes, ponds, reservoirs, and rivers combined. Their combined shorelines nearly match those of the Atlantic and Pacific coasts of the United States. From their shores one can see lighthouses guarding dangerous rocky shorelines, oil tankers and container ships arriving from around the world, ore boats from industrial ports, and even cruise ships.

The prairie potholes of the High Plains states also have not engendered lake culture. They are too small, windblown, and barren of trees. Most are completely surrounded by privately owned ranches and farms, offering little to the negligible influx of tourists. The locals use them as swimming holes or places to hunt ducks or "dunk worms," the pejorative phrase for fishing for small fish in small ponds.

Lake culture tends to be incompatible with great wealth. For the rich, privacy, isolation, and security are more important than the sense of community. Climate is a factor too, since lake culture is most strongly felt as a seasonal rhythm. Wilderness areas don't foster lake culture because they are more about survival and immersion into wildness than family fun. Lakes in the Boundary Waters Canoe Area in northeastern Minnesota are typically not where families water-ski, play cards on picnic tables, eat rhubarb pie, and put ketchup on everything. Instead it's a land of trail mix, mosquito repellent, freeze-dried dinners, sore muscles, splendid isolation, and limited shoreline development.

Fishing is the most defining outdoor sport of lake culture. For Thoreau it was a truly mystical experience. In the moonlight, he used his long flaxen line to communicate "with mysterious nocturnal fishes . . . feeling a slight vibration along it, indicative of some life prowling about its extremity, of dull uncertain blundering purpose." He found it "very queer, especially in dark nights, when your thoughts had wandered to vast and cosmogonal themes in other spheres, to feel this faint jerk, which came to interrupt your dreams and link you to Nature again."[26]

You can tell a lot about a culture by the type of fishing it does. Marlin, sailfish, and shark prowl the open sea. They suggest an oceanographic safari whose participants seek to escape the terrestrial realm and fish for combat sport. Trout, char, and grayling are found most often in cool, turbulent streams with pools and riffles, typical of forested hard-rock glacial terrains. This type of fishing supposes moderate income because of the expense of fly-fishing rods, hand-tied flies, perhaps wading boots, and overnight accommodations for nonlocals. Sunfish, bass, and pickerel live practically everywhere within the ponds, lakes, and sluggish rivers of the glaciated fringe. This type of fishing requires few expenditures, limited travel time, and can be enjoyed by anyone with a fishing rod, bait, and a few hours to spare. Dedicated trout fishermen are known to fly from one continent to another in search of the perfect stream. No fisherman I know has ever spent big money in search of the perfect kettle.

The first written account of fishing as a sport dates to 1496, when English publisher Wynkyn de Worde printed the anonymous "A Treatyse of Fysshyng with an Angle." This is the source for the word "angling" as a synonym for recreational fishing. Ever since then, exaggerated accounts of fishing exploits, also known as fish tales, have been part of the allure. Most such stories are assumed to be epic lies, so there is no point in even trying to be honest. "For a fish to be literary," wrote the angling addict Ian Frazier, "it must be immense, moss-backed, storied; for it to attain the level of the classics, it had better be a whale."[27] This is as true for Longfellow's *Hiawatha*—in which the hero's canoe is pulled underwater by a sturgeon—as for Hemingway's *Old Man and the Sea*, in which an enormous marlin nearly drags Santiago to his death. Fish that big cannot live in kettle lakes, which makes lying about "the one that got away" all the more important.

The largest game fish characteristic of heartland kettles is the muskellunge, largest member of the pike family. Not surprisingly, the

official record muskie wasn't from a kettle at all, but from an artificial
reservoir known as the Chippewa Flowage in Wisconsin. Caught in
1949, it weighed sixty-nine pounds. The second largest weighed fifty-
four pounds and was caught in 1957 from Minnesota's Lake Winnibigoshish, one of the largest kettles in the world. Michigan's heaviest
muskie, at forty-nine pounds, was caught from Thornapple Lake in
2000. Larger fish, of course, are often alleged, the lies about them accepted matter-of-factly. John Steinbeck, being an outsider to lake
culture, remarked in *Travels with Charley* that he had "no desire to
latch onto a monster symbol of fate and prove my manhood in titanic
piscine war." For him, fishing was catching "a couple of cooperative
fish of frying size."

One of my earliest memories of angling is of when I was a small
child fishing off the dock with a cane pole, a few yards of fishing line,
and a hook baited with a single kernel of sweet corn. Within seconds
a circle of sunfish appeared. Thinking back on this scene after fifty
years, I realize that those willing to bite were either the biggest, which
seemed to have bullied others out of the way, or the smallest, which
seemed to have been the most careless. Without knowing it, I had
been an agent of natural selection for midsized fish.

Fishing in large kettles can be more fun than in bedrock lakes of
comparable size because the chaotic terrain associated with ice stagnation continues under the water, albeit in a form subdued by the blanket
of sediment. Deep holes, weird drop-offs, almost-islands, and shallow
shoals are common, and seemingly random in their occurrence.


THE FREEZING OF THICK LAYERS of ice on lakes provides the setting for winter lake culture. Ice-fishing villages are northern utopian
communities where anyone can homestead a different patch of ice
each winter. Making a claim requires little more than skidding onto
the ice a small structure resembling a cross between an outhouse and
a one-room shack. From inside, people fish through holes cut in the
ice while warming themselves by stoves. Some also play poker, watch

television, and drink in the middle of the day. A few women participate in these communities, but it's generally a guy sort of thing. Ice-
fishing houses are the perfect social equalizers because the "land"
holds no real estate value at all, the ultimate in being house rich, land
poor. There are neither ghettos nor gated communities. Nor are there
real estate agents, taxes, liability insurance, easements, and neighbor-
hood covenants to deal with. And best of all, you can select new
neighbors each year without suffering capital gains or losses.

Other activities also take place on the ice. Cross-country skiing,
snowshoeing, and dog mushing create a weblike network of trails.
Movements of cars, trucks, and snowmobiles establish temporary
one-lane roads that are straight as a beeline and with no stop signs,
traffic lights, or shoulders. The normal rules of the road don't apply.
As teenagers, we used to race our cars as fast as possible on ice where
the wind had swept away the snow, with no fear of being stopped by
the police. For fun we would get up some speed, crank the wheel
sharply, slam on the brakes, and see how many times the car would
spin. I recall the record was eleven. Using a long rope, we would also
tow each other around on skis, sleds, or sheets of cardboard. One year
I put metal studs on my bicycle tires and rode all over.

Under the right circumstances, ice-skating can be perfect on a flat
sheet of lake ice that is orders of magnitude larger than any rink.
Writer Sigurd Olson described "seven miles of perfect skating,
something to dream about in years to come."[28] Ice-sailing is also
popular when conditions are right. The metal runners of iceboats
make friction negligible, and there is no chop or swell, allowing one
to sail close to the speed of the wind and much faster than on liquid
water, easily up to seventy-five miles per hour. A low-tech version of
ice-sailing with a parachute, rope, and sled can yield hours of fun.

During ice carnivals, the celebrants build elaborate castles and
cathedrals. Blocks of ice are quarried at no cost from the surface for
wannabe architects, then mortared together with liquid water. Giant
statues, especially of Paul Bunyan, are sculpted with chain saws and
axes. And as with Buddhist sand mandalas, there is little sadness

when these masterpieces melt away. Their impermanence is part of their charm.

Family lake culture, so much a part of the lives of contemporary Americans, is practically absent from our nineteenth-century literature. Walt Whitman grew up among kettle holes and ponds on Long Island.[29] Nathaniel Hawthorne and Henry Wadsworth Longfellow spent their childhood summers in the lake district of western Maine, in Raymond and Hiram, respectively, where kettles, partial kettles, and small bedrock lakes abound. James Fenimore Cooper grew up on the southern end of a quasi Finger Lake called Lake Otsego, which was dammed up by a kettle moraine to the south. Kettles dot the landscapes near Hartford, Connecticut, and Elmira, New York, where Mark Twain lived as an adult. To my knowledge none of these giants of American literature called attention to small lakes and ponds in their work.

Within the twentieth century, the direct link between literary culture and kettle lakes remained weak, even where they were most concentrated, largely because they tend to be pleasant, accessible, and uninspiring, rather than remote and dramatic.[30] Though small Minnesota lakes found their literary champion in Sigurd Olson, his inspiration came from the rocky lakes of the Canadian Shield, rather than the softer, more domesticated kettles to the south. New England poet and essayist Donald Hall was inspired by a kettle named Eagle Pond, though very few of his readers—and perhaps not even the writer himself—knew of its origin. Scandinavian author Vilhelm Moberg set his immigrant tale *The Last Letter Home* in classic kettle lake country, but he never found a large audience. Garrison Keillor's 1985 novel *Lake Wobegon Days* was set on the shore of a kettle lake, about which he says very little.

The indirect link between twentieth-century literary culture and kettle lakes, however, was extraordinary, coming almost entirely through Thoreau's *Walden; or, Life in the Woods.* Perhaps this book's literary success has deterred others from trying to copy its setting or purpose. But away from small lakes, the ideas about our relationship to nature expressed in *Walden* fueled writers of the twentieth century like gasoline thrown on a fire. Robert Frost greatly admired the book, which helped inspire his poetry. E. B. White wrote that "*Walden* is one of the first of the vitamin-enriched American dishes."[31] It inspired the works of Edward Abbey, Wendell Berry, Henry Beston, Rachel Carson, Loren Eisley, Annie Dillard, Joseph Wood Krutch, Robert Finch, Aldo Leopold, Sinclair Lewis, Barry Lopez, Bill McKibben, John Hanson Mitchell, Michael Pollan, Edwin Way Teale, Robert Pirsig, Roger Tory Peterson, Chet Raymo, Terry Tempest Williams, E. O. Wilson, Ann Zwinger, and countless others. These authors wrote about the desert, crashing surf, tidal marshes, sandy farmsteads, arctic tundra, bucolic green hills, majestic mountains, rivers, creeks, badlands, and more, yet each one expressed ideas inspired by a humble kettle pond.[32]

Garrison Keillor has also had a strong indirect influence on lake culture through *A Prairie Home Companion*, the best-known radio variety show in America.[33] The centerpiece for each show is a meandering monologue called "The News from Lake Wobegon," his archetypical small town on the edge of the prairie.[34] Though many of Keillor's monologues mention the lake, it's never the main focus. Yet it is always there in the background, giving a strong sense of place for the human dramas and adding a touch of background beauty to the pathos and the mundane. Keillor's oblique, passing, even shadowy treatment of the lake is perfect. It has allowed humble Lake Wobegon to remain humble and somewhat anonymous, meaning that it can represent in myth any one of the tens of thousands of unpretentious lakes and ponds in the glaciated fringe. Each of us with a favorite lake can think of it as a personal place where, ironically, woe is gone.

As the literary significance of Lake Wobegon has grown in the last thirty-five years, its actual importance in the fiction of its creator has receded. In each book Keillor seems to say less and less about the lake itself. In a 2001 photo essay, *In Search of Lake Wobegon*, Keillor finally tipped his hand, writing that the most "Wobegonic" town of all was Holdingford, Minnesota.[35] I looked into his suggestion. Holdingford is located on flat, open farmland more than two miles from the nearest natural body of water. The nearby town of Avon, however, is nestled against Middle Spunk Lake, one of the largest kettles within a local cluster. With pine forest on one side and prairie on the other, it looks far more "Wobegonic" to me. That's the kind of place I'd like to visit.

# 8.

# How Lakes Work

LIMNOLOGY, DERIVED FROM THE GREEK *limne*, for "marsh," is the official name for lake science. Scientifically, it's the fusion of geology, zoology, botany, climatology, chemistry, physics, and mathematical modeling applied to freshwater environments. Limnology is very similar to its sister discipline of oceanography, but is not as well known because the discipline is younger, less adventurous, and more intimate. Oceanographers work as teams of specialists with big budgets on enormous bodies of water. Limnologists are usually generalists who work on small lakes and who are expected to do it all on a shoestring.[1]

The scientific study of lakes began near the turn of the twentieth century in the European Alps at Lake Geneva, Switzerland, a glacially scoured, hard-rock, moraine-dammed lake of the sort that can be found in mountain ranges the world over. There François-Alphonse Forel, a professor of medicine at the University of Lausanne, summarized his life's work in three large volumes between 1892 and 1904. North American limnology began in the same decade, but in a very different place: the sandy kettle moraine country near Madison, Wisconsin. In 1895, prompted by a research paper on the plankton of European lakes, Edward Birge, instructor of natural history at the University of Wisconsin, published the first of his many scientific papers on Lake Mendota, which fronts the campus and is nestled within a prominent kettle moraine. In 1908 he teamed up with Chauncey Juday, who had recently arrived from the Wisconsin Geological and Natural History Survey. Combining their expertise in zooplankton and water chemistry, Birge and Juday began publishing dozens of papers, many comparing field observations from different lakes across the state. During the next several decades, they collaborated with other specialists, setting a precedent for limnology as an integrated subject.

Lake science caught on in other states as well. In 1895 Carl H. Eigenmann of Indiana University founded a lake research station on Lake Wawasee, a large, ragged kettle southeast of Chicago. David S. Kellicot of Ohio State University founded that state's Lake Erie laboratory one year later. Beginning in 1908, James G. Needham taught a course on general limnology at Cornell University. Paul Welch began teaching one in 1920 at the University of Michigan. His experience with the state's many kettles led to the publication of the first limnology textbook in 1935, which facilitated its instruction elsewhere.[2] Meanwhile, in Branford, Connecticut, G. Evelyn Hutchinson of Yale University began using nearby Linsley Pond as a natural laboratory for aquatic systems. An Englishman, he had arrived from South Africa in 1928, bringing with him the holistic European approach to natural science.

Limnology finally emerged as a recognizable discipline in 1936, when scientists from across the country gathered to form the Limnological Society of America.[3] By 1948 they realized that they shared so much in common with their saltwater brethren that they merged to form the American Society of Limnology and Oceanography. The current gold-standard textbook in the field, the late Robert Wetzel's *Limnology: Lake and River Ecosystems*, third edition, defines limnology as "the structural and functional interrelationships of organisms of inland waters as they are affected by their dynamic physical, chemical, and biotic environments."[4]

The 1957 publication of Hutchinson's *Treatise on Limnology*, the first of four volumes, unified the discipline. He is generally considered to be the father of modern limnology, not because he was the first to publish important ideas, but because he spent much of his later professional life completing an authoritative quantitative synthesis of the flow of energy and matter between the terrestrial, aquatic, and atmospheric realms, with an emphasis on different levels of the food web. From my perspective, his most important contribution was to recognize kettle ponds—being small, isolated, and self-contained—as ideal places to investigate ecosystems because they were far simpler than larger lakes or flowing aquatic settings such as streams, estuaries, and the sea. According to Wetzel, the use of kettle lakes as microcosms for bigger and messier ecosystems helped produce a "comprehensive understanding that is unrivaled by any other area of ecology."[5]

One of Hutchinson's early students, Howard Thomas Odum, generalized his kettle pond experience, first to larger ecosystems, then to biomes, then to human impacts on biomes, and finally to global bioeconomics and bioengineering. His textbook *Fundamentals of Ecology* (first edition, 1953)—coauthored with his brother, Eugene—underpinned the scientific basis for the rising tide of environmentalism during the second half of the twentieth century and, more recently, for the concept of ecosystem services, which places a dollar value on the services natural processes contribute to human well-being.[6] Edward Deevey Jr., another of Hutchinson's students, helped launch the American study of past climates, using the fossil archive contained within the muck. Building on pioneering studies in Sweden, he used the pollen record from Linsley Pond to reconstruct the postglacial succession of ecosystems in northeastern North America and the paleoclimatic changes that brought them into being.[7] Finally, the isolation of small kettles and their characteristic occurrence within highly permeable, fairly uniform, and easily drilled deposits of sand and gravel made them ideal natural laboratories for the pioneering studies of groundwater hydrology as a distinct scientific discipline.

Three limnology programs have become leaders, all at land-grant universities in states where kettle lakes are signature landforms. At the University of Wisconsin, the Hasler Center for Limnology lies on the shore of Lake Mendota, where American limnology began. To the north is the university's Trout Lake field station, location of one of the most well-studied kettles in the world. At the University of Minnesota, the Limnological Research Center, under the direction of Herbert E. Wright, follows a long tradition of studying the paleoecology of lakes based on their sediment cores. In Michigan, the Cooperative Institute for Limnology and Ecosystems Research is jointly managed by the University of Michigan and Michigan State. This program befits the state most surrounded by freshwater seas.

HENRY DAVID THOREAU was doing limnology at Walden Pond more than half a century before the first academic articles by Forel and Birge.[8] The entire text of *Walden* is laced with lake observations. Its only illustration is an accurate bathymetric (depth measurement) map drawn from soundings he made by dropping a weighted cod line through ice holes chopped with an ax. Two chapters, "The Ponds" and "The Pond in Winter," focus on how Walden Pond, like Linsley Pond, works as a unified system through all four seasons. Thoreau's descriptive and comparative scientific work at Walden Pond is the true beginning of American limnology as an adjunct of natural history.

Thoreau paid careful attention to the physical lake basin, properly interpreting the origin of the pond and the distribution of its geological materials. He was fascinated by the optical properties of the pond in its solid, liquid, and gaseous phases. He noted how the water changed colors depending on the time of year or time of day, the perspective of distance, at what angle one looked, and the stability of the atmosphere. This involves the optics of solar rays and the diffraction, refraction, absorption, and scattering of its spectrum. He understood that the clarity of the water was due to groundwater seepage and proved this with water temperature data. He theorized about the atmospheric conditions during which the pond cooled most rapidly. He was the first to publish on the thermal stratification of lakes, a defining characteristic of northern lakes and an important factor in controlling the hidden movements of water throughout the year.

As a pioneering aquatic ecologist, he commented on the wildlife and its habits, especially the fish: perch, pouts, shiners, chivins, roach, trout, bream, eels, and at least three distinct kinds of pike or pickerel. Regarding the flow of energy and nutrient through food webs, he wrote, "The perch swallows the grub-worm, the pickerel swallows the perch, and the fisherman swallows the pickerel; and so all the chinks in the scale of being are filled."[9] He even made the link between lake basin topography and biological productivity, noting that the larger and shallower Flint's Pond was "more fertile in fish" and "not remarkably pure" when compared with Walden Pond. This relationship holds today because the average depth of a pond is one of the most critical factors affecting its thermal behavior and response to nutrient inputs.

LAKES ARE FINELY TUNED MACHINES fueled by five separate manifestations of solar power. Scattering of light, which is electromagnetic radiation in the visible spectrum, by water molecules is responsible for the timeless blue color of lakes and ponds. When the sun strikes at a low angle, as it always does during sunrise and sunset, light is reflected by the water surface. Such reflections are responsible for the silvery, yellowish, and reddish sheens from placid water, the strobelike pulses from uniform waves and ripples, and the sparkles from transient turbulence caused by gusts of wind. When the sun is high overhead, however, photons of light penetrate deeply downward into clear water through a thicket of water molecules until they are scattered, giving rise to the blue and aquamarine colors we enjoy so much.

In ultraclear lakes, light can penetrate to a depth of about sixty feet, though only half that is the normal limit for easy-to-see conditions. Brightness diminishes with depth, as does the intensity of shadows. Continuous shadows are cast by a bank of trees above the shore, mottled ones by lily pads, and striped ones by submerged reeds. A glaze of ice is a transparent lens, a dusting of snow a translucent film, a coating of duckweed a fine-meshed gauze. Particles suspended in the water dim the light at depth much faster and change the color of the water. The turquoise hues of some mountain lakes,

*Sunrise over Lake Plantagenet, Minnesota.*

the pinks of equatorial rift lakes, the latte color of lakes fed by muddy streams, the pale greens of pools, and the tea color of bog ponds are due to small particles of lime, plankton, clay, algae, and dissolved organic acids, respectively.

In contrast to the blue-green wavelengths that scatter to give clear lakes their default coloration, red and infrared wavelengths are absorbed by the water molecules and are converted to heat, mostly within the uppermost few feet. This would be a pretty simple phenomenon were it not for the weird thermal behavior of water. Practically every natural substance on Earth is heavier (denser) as a solid than as a liquid. Not so with $H_2O$: Ice crystals in water float instead of sinking. Even weirder is the fact that liquid water reaches its maximum density not when it is coldest, but at thirty-nine degrees Fahrenheit, its so-called critical temperature, seven degrees above its freezing point. As a result, all lakes deeper than about fifteen feet develop different thermal layers in summer and winter that, though invisible, profoundly influence the local ecology.[10]

Kettles along the southern part of the glaciated fringe from Illinois to eastern Pennsylvania, and those influenced by maritime conditions along the Atlantic coast, don't freeze hard enough in the winter to develop a lasting layer of ice. Their winter waters are either uniformly mixed or have a layer of cool water over cold. During spring and summer, a discrete layer of wind-stirred, solar-heated water develops and floats above whatever cold water is left over from the previous winter.[11] As autumn progresses, water chilled near the surface cools and descends via slow-moving eddies during what's called autumn turnover. If the weather is cold enough, this process continues until the whole lake is homogenized at a temperature of thirty-nine degrees. If cooled below that temperature under calm conditions, the surface water begins to expand in volume, making it less dense and causing it to float above the water that went down earlier. Inevitably, these surface layers of near-freezing water are transient, destroyed when the wind picks up again.

The majority of kettles to the north and in the continental interior develop thick ice during the winter, especially in the heartland.[12] For most of the year, they behave as do more southerly lakes, developing a layer of warm water in summer, overturning in the fall, homogenizing at the critical temperature, and developing a transient layer of chilled water on the surface. But on the first bitterly cold night, especially if the wind is blowing gently but steadily, the top layer of water can be chilled to a temperature below the freezing point. When the surface is becalmed, usually before dawn, a glaze of ice will quickly freeze. This shields the liquid lake away from the turbulent atmosphere, allowing a stable layer of near-freezing water to form beneath the ice. As winter drags on, this layer thickens above the merely cold water occupying most of the lake volume. The layer thickens even more in late winter when slush-cold meltwater leaks downward, through fractures in the thawing ice. When spring arrives, however, this near-freezing surface water warms to the critical temperature, descends, and mixes the lake to a homogenous temperature once again. This is called spring turnover.

Water can hold more heat energy than practically any other

substance, five times more than an equivalent volume of rock. Also, the melting of ice and the evaporation of water sop up huge amounts of heat energy and help keep lake water within a fairly limited range of temperatures; this thermally buffers the temperature of lake water. On blazing-hot or bone-chilling days, heat and cold that might otherwise kill fish are instead dissipated. Water's thermal inertia also buffers the local climate. Lake air is colder in the spring and warmer in the fall for this reason as well.

The weird thermal behavior of water, the tendency of kettles to "stratify" in stable layers during summer and winter, and the requirement that they mix completely during spring and fall mean that their underwater environments are more moody and volatile than those of shallower or more southerly bodies of water. Slow currents of water with different amounts of heat, nutrient, and dissolved chemicals move up, down, and sideways at various times. The irregular subsurface shapes of most kettles amplify and attenuate these fluid motions, to which fish and other organisms respond. The best fishermen are those who understand how these hidden currents move and what they carry. Paul Nelson, who writes about fishing for northern Minnesota's *Bemidji Pioneer*, is a remarkably good lake scientist, albeit one without scientific training and credentials. Occasionally my mother will clip one of his columns out of the paper and send it my way in New England.

Another manifestation of solar power influencing kettle lakes is the wind, the flow of air caused by its differential heating. At the bottom of every moving air mass is a turbulent boundary layer responsible for causing the sway of plants on land and a choppy surface on the water. Waves appear more quickly and move faster on lakes than on the sea because the water is less dense, especially in late summer. When wind shears over the roughened surface of a lake, some of its momentum is transferred to the water, creating periodic waves that migrate away from their source. In the process, the kinetic energy contained within a volume of moving air is transformed into the wave energy moving over the lake surface, which is transformed by friction into the mechanical energy of splash and swash against the line of the shore. The strength of waves is controlled by several factors, including the distance over open water in the direction of the wind, the duration of squalls and storms, the topography of the shoreline, and the contour of the lake. Small kettles hunkered low in the woods and with steep sides experience much smaller waves less frequently than do more exposed bodies of water the same size. The rise and fall of water levels common to elongate lakes called seiche waves—similar to tides but caused by a different mechanism—are insignificant in kettles, owing to their blotchy shapes and small sizes. In a nutshell, classic kettles have the internal and seasonal water movements typical of deep lakes, but the shoreline behaviors more characteristic of ponds. Each is a unique hybrid of behaviors.

LAKES ARE ALSO POWERED by precipitation—a fourth manifestation of solar power—in ways far beyond the obvious fluctuations in stream flow and lake level. Precipitation sets up important, long-lasting, and invisible changes in the potential energy of individual lake systems through the storage of snow on the land and raising and lowering the groundwater mound between and near them. Raising the mound with an infusion of melting snow and rain trickling downward inaugurates a time-delayed pulse of aquifer flow into lakes. Flattening the mound relaxes the pressure and reduces the surface gradient of the water table, diminishing flow. These feedbacks help maintain a system of springs and seepages that freshens kettle lakes continuously.[13]

Groundwater drains slowly beneath the land, moving mostly in a horizontal direction with sluggish flow rates ranging between a few inches and a few hundred feet per day, depending on the granular texture of the aquifer. In coarse gravel, the lake may rise and fall in close sympathy with infusions. In finer-grained sand, and when there is a great distance between the crest of a groundwater mound and the shoreline of a lake, it may take several years for a pulse of water to reach the lake and raise its level. Thus, the rise and fall of the shore-line of kettles is usually out of phase with the wet years and droughts that cause it. Thoreau found these fluctuations, "whether periodical or not," remarkable, because each required "many years for its accomplishment."[14] Another outcome of the sluggish but steady flow of groundwater is the fairly long time needed for an average molecule of water to flush through the lake system, far longer than for a lake of comparable size fed by a stream.[15] This makes kettles slower to respond to an on-land pollution event, but causes them to suffer longer.

The movement of water through aquifers also influences the chemical composition of the water, especially salts, which arrive as dissolved, electrically charged atoms and molecules called ions.[16] Calcium, magnesium, sodium, and potassium are positive ions. Carbonate, bicarbonate, sulphate, and chloride are negative ions.[17] Lakes with high levels of dissolved solids and rich in salts are said to be "hard." They are much less prone to damage from acid rain because their natural alkalinity buffers incoming acid. They also are less prone to nutrient pollution because the salts react with phosphorus to form insoluble compounds that sink to the bottom, rendering them unavailable for algae and weeds. The so-called freshwater in kettles is actually highly variable. Those draining through crushed-granite sand in a moist climate like that of central New Hampshire have very little salt. Those draining through crushed marine shale in a dry climate are saltier than the ocean, commonly with white crusts of precipitated brine. Herring Pond in Wellfleet, Massachusetts, is saltier than normal because droplets blow in from the foaming surf less than a mile to the east. Pleasant Lake in Erie County, Pennsylvania, is remarkably clear because its spring-fed water has drained through crushed limestone.

More generally, the volume and chemical properties of water within a kettle basin at any time depend entirely on how its plumbing system communicates with three other water reservoirs: the atmosphere, groundwater, and stream flow.[18] If evaporation goes up while everything else stays the same, there will be less water and it will be saltier. If the outlet runoff goes up, there will be less water but its saltiness will be unchanged. If a lake is fed principally by surface rain, it will have few dissolved solids and little pollution, assuming the air quality is fine. If the same lake is fed by streams, it will have more dissolved solids and more pollution. If fed by groundwater, it will have the maximum amount of dissolved solids but less pollution. In effect, the mass balance and chemical balance for every lake is not only different but always changing. All in all, spring-fed kettles are less flashy but more quirky and touchy than stream-fed lakes of the same size.

THE FINAL MANIFESTATION OF SOLAR POWER acting on lakes is photosynthesis, the conversion of sunlight to the chemical energy of food. Physically the process begins when a photon strips an electron from an atom of hydrogen within water. Green plants then take the electrically imbalanced hydrogen and combine it with carbon, nitrogen, and phosphorus to make plant tissue, the ecological bedrock of the food chain.[19] Hydrogen and carbon alone create the simplest organic molecules such as methane, aka swamp gas. Combining these elements with oxygen gives rise to the greater complexity of sugars, starches, and lipids (fats and oils). Nitrogen and a smidge of sulfur yields proteins, the building blocks of life. Phosphorus yields a critical molecule abbreviated as ATP, which is essential for all animal respiration.[20] Many other elements known as micronutrients—notably zinc, manganese, potassium, boron, molybdenum, vanadium, iron, fluorine, strontium, and copper—are also necessary for special-purpose enzymes and catalysts.

Submerged aquatic plants must get all of their nutrients directly from the lake water. There's no shortage of hydrogen, because every molecule of water contains it. Carbon and nitrogen must come from the atmosphere, dissolving either directly via gas diffusion or indirectly within organic compounds dissolved in inflowing water.[21] Phosphorus is the only major nutrient that does not cycle through ecosystems in a gaseous phase. Instead it enters freshwater either as a dissolved mineral ion, as dissolved organic matter, or as a hitchhiker attached to particles in runoff. Estuaries and oceans are seldom starved for phosphorus because that's where the excess from land flows until it's sequestered within marine sedimentary rock. In the sea, nitrogen is usually the biogeochemical throttle for the biological productivity of the food chain.[22] But with freshwater aquatic systems in general and with kettles in particular, phosphorus is the critical limiting nutrient.[23] It runs the supply side of the aquatic economy.

Charging a battery with electricity takes place by increasing its chemical potential energy. Something similar happens when an ecosystem is charged by photosynthesis. With a battery, the energy is stored by the rearrangement of salts and metals. With an ecosystem, the energy is stored by the rearrangement of nutrients into progressively higher concentrations of energy: gas, starches, sugars, proteins, and lipids. A battery will remain fully charged until its terminals are connected to a circuit. An ecosystem will remain charged until oxygen becomes available to draw that energy away during a process called respiration. Oxygen runs the consumption, or demand, side of the aquatic economy.

Though each molecule of water contains an atom of oxygen, this form cannot be used by the processes of life. Bioavailable oxygen must be dissolved from a gas. This process happens slowly when the water is quiet, much faster when it is bubbled by mechanical turbulence via wave and stream action. This is why lakes and ponds with algae problems often have fountains to aerate the water. Boat propellers and splashing children oxygenate water as well. The most important source of dissolved oxygen in many lakes, however, is photosynthesis, which fortuitously releases the gas as a waste byproduct. Tiny bubbles appear on the edges of submerged leaves, rise upward, and then usually are dissolved into the water before reaching the surface. (Thoreau was fascinated by such bubbles that got trapped in winter ice.) Aquatic creatures, which get what oxygen they need either directly through their skin or gills, must compete for this substance with other oxygen-consuming chemical processes, especially the bacterial decay of organic matter. The natural rusting of iron-bearing minerals also sops up oxygen.

At any moment, the amount of oxygen in a parcel of water depends on its location, temperature, and pressure. Water near a choppy lake surface, especially if there are whitecaps or a turbulent incoming stream, is usually rich in oxygen due to diffusion from the atmosphere. Cold water can hold more oxygen than warm water. Deep water under pressure holds more than shallow water at lower pressure.[24] Hence, the thermal layers that form in lakes during summer and winter, and the currents that overturn them during fall and spring, either concentrate, dilute, or blend the oxygen supply, which, in turn, controls what happens biologically.

A lake fed with a tiny bit of phosphorus is nearly sterile. This situation occurs most often in boggy lakes, where the high concentration of dissolved organic acids or iron creates a tea that binds the phosphorus. These ultralean lakes are eerily quiet and with so little nutrient that carnivorous plants are common along their edges. A lake fed just a little more phosphorus is lean, creating an ecosystem adapted to an intense competition for nutrients and with an abundance of dissolved oxygen. In such a circumstance, the decomposition of detritus by oxygenated respiration outpaces its production, giving rise to lake bottoms dominated by sand and gravel, those that species like trout prefer. Lakes fed plenty of phosphorus are biologically richer. They have more weeds, are slightly less clear owing to a higher concentration of suspended algae, and have more animal biomass. Bass, pike, and sunfish are typical. Catfish and carp thrive in lakes that are loaded with phosphorus, which causes much more organic matter to be produced than can be decayed, given the oxygen supply available.[25]

WHEN LAKES ARE ENORMOUS, like, for example, Lake Superior, the best way to appreciate their geography is with the traditional checkerboard grid of latitude and longitude. When lakes are long and ribbon shaped, like New York's Finger Lakes, they are best understood as wide lines. Lakes in stream valleys dammed up as artificial reservoirs, like those of the Tennessee Valley Authority or the canyons of the Southwest, are most easily visualized with a hierarchical geography similar to that of a branching tree, in which tiny triangular bays join to form bigger bays, then even bigger bays, and so forth. Simplest of all lake geographies is that of a classic kettle lake formed as a sand sinkhole where an ice block once was. This is a circular geography. The same is true for each bay within larger composite kettles formed by multiple ice blocks.

The most logical place to begin describing kettle lake environments is the bull's-eye center of the lake. In small kettles this coincides with the deepest part of the lake. Moving out in all directions from the center are a series of rays that cross a series of concentric zones. The innermost is the deepwater zone, where three layers of water usually lie one above the other, and where there are no rooted plants. The next zone outward is that of shallow water. Rooted weeds rise up from the bottom because the water isn't too deep for photosynthesis. Next in sequence is the transitional zone, which takes the form of a bouldery shore where exposure to wave energy is high, a narrow sand beach where it is intermediate, and a broad wetland where the physical energy is low. The zone farthest from the lake center is the terrestrial realm, which can be forest, prairie, urban-suburban, or developed in various ways for recreational purposes. Though not technically part of the lake, the terrestrial zone is—chemically, hydrologically, biologically, and culturally—part of the lake system. It extends from the top of the beach or saturated soil to the groundwater divide separating the lake in question from an adjacent body of water or flowing stream.

In small classic kettles below a flat outwash plain, the boundaries between the deepwater, shallow water, transitional, and terrestrial zones, and the outermost perimeter of the groundwater mound, approximate the shape of a circle or an ellipse. With larger and more irregular kettles, the same sequence of environments will be encountered, but the boundaries will be more irregular. The shoreline wiggles around in one of three modes: eroded headland, accreted beach, or vegetated wetland. Though this is the most visible perimeter of the system, the most important one for the chemical health of a kettle is the invisible crest of the groundwater divide. Every drop of water and every molecule of contamination, unless adsorbed or evaporated, has no choice but to infiltrate downward to the water table, then sideways to the lake. In hilly ice stagnation terrain the divide usually approximates the local topographic divide. In flat terrain, it must be mapped out by subsurface investigations, usually the depth of water in a population of water supply wells.

Within each of the lake system's concentric zones is a fairly predictable series of layers stacked one above the other. Present beneath all zones is a layer of water-saturated earth usually composed of either sandy gravel or till. No one has seen this layer, but geologists, hydrologists, and pollen scientists are certain it's there based on field observations, countless well borings, and sampling tubes pushed all the way through the lake sediments, respectively.[26] This quiet, deep, and stable layer—where the most dramatic action is the slow flow of groundwater moving through the voids—is the true bottom of the lake system. Little happens except for the slow underground flow of water through billions of voids. There's plenty of material variation, however, perhaps a lens of clay that will block the flow or a lens of clean gravel that will enhance it. Remarkably, no visible sign of life is present, though invisible bacteria are ubiquitous.

Present above all zones is the atmosphere. Sometimes it appears still. During severe storms it's violent and chaotic. When the prevailing winds are blowing, there is a steady flow of air across the entire system, bringing in heat or cold, water or drought, and aerosols from natural and man-made sources. When they are still, local breezes caused by the uneven heating of land and water move in and out like breath from a lung. The shapes of the land either strengthen or diminish air currents, especially at an isthmus between bays or on the downwind side of a broad stretch of open water.

A dark-colored organic horizon of some sort is present in every zone except on narrow beaches. On dry land, it is the familiar topsoil, through which rain and runoff infiltrate, as if through a coffee filter, dissolving and depositing chemicals. In the wetland zone, the organic horizon is not a flow-through system like that of drained soil. Instead, it is a reactive interface between the oxygen-starved soil at greater depth and the seasonally aerated soil at the surface. Weird colors—mottled bright reddish oranges and greenish grays—and barnlike smells manifest the important microbial work being done and the chemical exchanges that keep the lake system healthy. In the zone of shallow water, the organic layer consists of fibrous detritus dominated by shredded and macerated weed tissue resembling felt soaked in dark pudding. In the deepwater zone, the organic layer is muck with the consistency of a stiff gel and the color of grayish molasses.

In deep lakes, the muck is usually enriched during late summer when an excess of detritus falls to the bottom. During the winter, this excess is recycled back into the lake by bacteria, which liberate chemicals—calcium, iron, sulfur, nitrogen, and phosphorus—that dissolve back into the cold, stable bottom water. When the lake experiences spring turnover, the upflow of nutrient-laden deep water sets the stage for a later surge of biological productivity coinciding with summer warmth. Also moving toward the surface is carbon dioxide, the waste gas produced by respiration. It fizzes out and is lost to the atmosphere as the pressure drops in the same way a carbonated beverage becomes bubbly when the cap is removed. Also rising up are mineral salts that arrived through winter groundwater seepage. During the autumn turnover, cold surface water, often richer in oxygen, descends to recharge lake bottom life.

Beneath the terrestrial zone, and only there, is a vertical layer where earth materials are neither organic nor saturated, except on rare occasions. There, beneath cottages, cabins, camps, sheds, houses, driveways, and urban areas, the water from garden hoses, septic tanks, drains, emptied buckets, snowmelt, and rain soaks straight downward. If the soil below the surface is dry, the water will be sponged up. If it's saturated, the water will trickle down to saturate the porous material between the soil and the water table. Only when everything is sopping wet will downward-trickling water recharge the aquifer and raise the water table. This usually happens in early spring, when snowmelt and drenching seasonal rains combine forces at a time when the uptake of water by vegetation is on hold.

The presence or absence of snow and ice on the ground makes a huge difference with what happens in the unsaturated mineral layer. If a rainstorm happens on frozen but snow-free ground, the infiltrating water will freeze, creating an impermeable barrier that will shunt much of the spring snowmelt and rain sideways as runoff, which usually causes more pollution to enter lakes. As a result, the groundwater mound is only weakly recharged. Conversely, a thick, early layer of snow on the land surface will insulate the soil enough to keep it from freezing deeply. During spring, infiltrating water from snowmelt will strongly recharge the groundwater mound, which allows clean, cold, filtered water to seep into lakes all summer long.

The sensitive toggle switch between surface runoff and groundwater recharge is one of the reasons why northern kettles are more temperamental than other lakes of similar size. Their habit of developing a thick cake of ice makes them even more touchy. If the ice remains thin and uncovered by snow, it can act like the windowpane in a greenhouse, allowing photosynthesis to continue and the water to become oxygenated. This transmission of light can also warm the water beneath the ice back up to its critical temperature, causing oxygenated water to sink to the bottom, influencing nutrient recycling. Alternatively, multiple freeze-thaw cycles during early winter can create enough slush to send the lake into premature darkness. Snow, of course, will also darken the lake, but it is free to sublimate and blow away. Because kettles trap cold pockets of air, the ice is usually thicker and the snow more long lasting. Thoreau likened the freeze-up transition of Walden to that of a hibernating marmot, which "closes its eye-lids and becomes dormant for three months or more."[27]

When kettles wake up in the spring, their icy shorelines experience the most dramatic action of the year. By the end of winter, the cake of ice on a typical kettle in the north woods can be up to three feet thick. Thick ice results from clear-weather cold snaps before the receipt of an insulating snow cover, the conversion of deep snow to frozen slush, or unusually cold conditions. Near the base of the ice, the temperature is, by necessity, very near the freezing point because it's in contact with

*Close-up of ice on a kettle pond in Rhode Island.*

liquid water. At the surface, however, the ice can be tens of degrees below zero. Like most solids, ice is a brittle material that expands when it warms and contracts when it cools, which is why it makes so much noise.[28] When a thick sheet of unbroken ice abruptly warms (but doesn't melt) in the spring, it expands horizontally, pushing and thrusting against the shore, breaking every dock in its path, and aligning stones on the shore into pavements like those that so puzzled Thoreau. If the ice cover breaks up mechanically due to a rise in the water level beneath the ice, sheets and floes can be thrust against the bank, pressing, gouging, scraping, and smashing almost to the point of glacial action. Floes created during melt thinning usually ride high on high water. When blown hard into the bank by strong spring winds, they cause great damage, especially to low-growing trees, many of which bear battle scars or, to use Thoreau's words, are shorn away.

The complexity of even a small lake system is astonishing. In winter, the deepwater zone has seven important layers arranged in order of increasing density: air, snow, ice, near-freezing water, water near or at the critical temperature, muck, and saturated mineral sediment. This does not count the silt liner as a separate layer. Greater complexity arises when variations along the lake perimeter—exposed bluff, marsh, and beach—are taken into account. Even a small, seemingly simple kettle lake system can have dozens of environmental compartments experiencing hundreds of different behaviors during any given year.

The final challenge in understanding the kettle lake environment involves the seasonal offset between the climate above the lake and that above the land. In spring, ice floes can still be present long after the crocuses and daffodils emerge. At that time, the air above the lake is colder than above the land, creating breezes that flow outward toward the land, especially on warm afternoons. These are the bone-chilling moist winds of April. In autumn, the breezes go the other way because the lake, having adsorbed heat all summer long, is now a source of warmth, causing the air to rise. These are the dry, delightful, pine-crisp days of September.[29]

The physical and chemical characteristics of each kettle lake provide the foundation for the more visible and living part of the lake system. Any attempt to maintain a healthy lake ecosystem must begin with an understanding of how lakes work. Saving a fish might require knowing an ion.

9.

# Habitats, Flora, and Fauna

KETTLE LAKES AND PONDS teem with life. Naturalist John Muir described the scene fronting his boyhood home on the kettle moraine of southeast Wisconsin. Fountain Lake was

> surrounded by low finely-modeled hills dotted with oak and hickory, and meadows full of grasses and sedges and many beautiful orchids and ferns. First there is a zone of green, shining rushes, and just beyond the rushes a zone of white and orange water-lilies fifty or sixty feet wide forming a magnificent border. On bright days, when the lake was rippled by a breeze, the lilies and sun-spangles danced together in radiant beauty.[1]

Woodland, meadow, lush herbs, rushes, lily pads, and open water—each band of life gives way to another, from the driest, highest, and outermost reaches of the land to the center of each lake or pond. In every place and at every time, all of these living things at every taxonomic level—microbes, herbs, insects, birds, fish, trees, and quadrupeds—are part of a single, living ecosystem. Each habitat within it matches one of the physical compartments of the kettle lake system.[2]

The food webs of classic kettles can be compared to the beautiful silken webs of orb-weaving spiders. The primary threads run outward from the center of the lake in all directions. Along these radial lines, biological energy flows back and forth between the deepwater, shallow-water, transitional, and terrestrial zones. For example,

*Cape Cod habitats, Brewster, Massachusetts.*

plankton from the center can feed a small fish that swims to shore, is caught by a raccoon, and is brought deep into the woods. From there, the organic carbon flows back to the lake as small particles washed during storms or, more commonly, as dissolved organic molecules. The secondary threads run parallel to the shoreline as a series of widening circles. Consider the bobcat that prowls the length of the beach in search of food, the northern pike that cruises the edge of the drop-off, or the near-shore current of plankton-rich water flowing steadily toward a bed of clams. As with the spider's surprisingly strong orb, the radial and concentric threads of food webs reinforce each other to keep the system stable and resilient.

A third dimension to the kettle food web flows vertically up and down. The excretions of flying birds when dissolved into the lake provide nutrients for the diatoms, which will sink to the bottom and be consumed by insect larvae that will float back up to hatch and feed the birds. In the center of the lake, the main action takes place at the water's surface. It is here that opportunity and danger are greatest. Photosynthesis is strongest, the water is warmest, and the visibility is highest. Anything heavier than air but lighter than water will reach the surface, either by falling from above or rising up from below. Organisms are drawn to the availability of biological energy like iron filings to a magnet, algae congregating where they find the most sunlight, zooplankton the most algae, small fish the most zooplankton, and predators where they can easily see their prey from both above and below. Carnivorous fish and otters rise to the surface to feed. Eagles and kingfishers dive from above. There is nowhere to hide. Nothing is fixed in position. Everything drifts on currents of water and air that are energized by the sun, or is propelled by muscular contractions that are also energized by the sun via photosynthesis.

When the wind is blowing, the interface between water and air is more challenging for an organism than either the air above or the water below. A choppy surface creates visual and physical confusion. The drag of the wind creates a sheetlike current of water against which an organism must work, or with which it gets a free ride. Waves move in circular orbits that cause bobbing motions. Whitecaps whip up a chaos of bubbles that shadow the water with transient shade.

The lake-sky interface is governed by the unusually high surface tension of water. Each molecule has two small atoms of hydrogen attached nearer to one side of an oxygen atom than the other: $H\text{-}O\text{-}H$. This makes the molecule electronically lopsided, giving the surface enough electrodynamic tension to support solid objects like water beetles. High surface tension also helps keep pollen and fallen insects on the surface, where they are easily visible and provide food for zooplankton and fish. This property, however, is a distinct disadvantage for moths and other insects, which once wetted are trapped because they cannot escape the molecular pull. Beyond the shoreline, water's high surface tension allows it to rise upward against gravity via capillary force, as with a towel dipped into water. This fringe of moisture provides a natural buffer against surface runoff by supporting lush vegetation, even during droughts on the prairie. Finally, water's high surface tension prevents shorelines from becoming foamy. Suds form only when the molecular tension is reduced by natural and artificial chemicals.

Water below the surface is a tempered milieu for life. Chaos diminishes. Noises are muted. Temperatures change slowly. Movement is slowed. Water's higher density relative to air allows heavy creatures like fish to hover using practically no energy at all because tiny adjustments in the size of their air bladders keep them neutrally buoyant. The liquid stiffness of water, known as its viscosity, helps tiny plankton stay at the same vertical level, wherever their preferences for heat, nutrient, or sunlight are best met. This property also explains why fast-swimming creatures such as fish, loons, and otters evolved sleek bodies. On land, the race may go to the swift, but underwater, it goes to the streamlined.[3]

In broad, deep lakes, single-celled algae are the dominant plants, equivalent to grass on the prairie or trees in the forest. These phytoplankton are the "very stuff of life," wrote naturalist Diana Muir. "Tiny leaves . . . turn their faces to the sun and create matter. It is a miracle no member of the animal kingdom can perform."[4] Zooplankton graze on these green specks like cows in a pasture. Single-celled plankton include paramecia, amoebas, and hydras. Multicellular versions are about the size of a sand grain and include rotifers, water fleas, and various invertebrates resembling either miniature clams or miniature pill bugs. Large zooplankton are the most important food sources for minnows and juvenile fish called fry, on which medium-sized carnivores like perch and sunfish feed. The largest piscine predators of open water come from the bass, walleye, and pike families.

Below the wind-stirred, food-rich layer of surface water is a twilight zone of colder, darker, quieter water. It remains still and stationary, except when it's being mixed during fall and spring turnovers. Through this medium falls a steady drizzle of dead organic matter—grains of pollen, diatoms, dead water fleas, seeds, stems, wood fragments, and algal cells. A few of these fragments are consumed on the way down, but most will eventually reach the bottom, sometimes after days of settling. Sinking more rapidly through this zone are dead fish and objects either lost or thrown overboard from boats.

At the base of the deepwater zone is the interface between water and muck, not quite liquid, not quite solid, but something in between.

That which is not eaten or dissolved will become part of the lake sediment archive, which consolidates and stiffens with depth. If oxygen is present, this dim world surges with microbial and animal life because there is so much food lying around. Bacteria dining on the muck are, in turn, fed upon by larger bacteria, and they by scavenging one-celled creatures, which are fed upon by larger creatures, most of which are insect larvae and various kinds of worms.

The most well-known creatures nourished by the muck are a large group of insects known as lake flies, mayflies, and damselflies. Like the cicadas of the forest, they spend most of their lives as larvae burrowing in the muck, eating near the bottom of the food chain. But when mature, they metamorphose into pupae, float slowly upward through the quiet water, and reach the surface, where they prepare to hatch. They show up mysteriously, like manna from below, eaten by all sorts of small carnivores, principally fish and insects. Those that survive hatch, as short-lived winged insects that fly away in an attempt to mate. Unless eaten by birds, bats, and dragonflies, they fall back to the water surface with their fertilized eggs, struggle momentarily, and then sink back to the dark bottom as detritus. There the eggs will hatch to begin life anew. Much of the wildlife we witness near the lake shoreline is actually fed from below. Biologically, the fetid, dank bottom of the lake is nearly as important as its top.

NEARER THE SHORE IS A ZONE of shallow water. On exposed, erosion-resistant shores, it is usually narrow, steep, and stony, a rarity for most kettles. In the more usual case, where the bluffs are composed of soft sand and gravel and have been exposed to a modicum of wave energy, this zone resembles that of a miniature continental shelf, planed off by waves during the slow drop in the water level of many lakes during the middle of the postglacial epoch, and its subsequent rise. Broad, shallow edges are particularly characteristic of those lakes that have shorelines that rise and fall at hydrological frequencies governed by the groundwater mound. Regardless of the zone's width, gentle currents of cold and warm water flow in all directions, depending on daily and seasonal thermal cycles and on currents set up by local and prevailing winds. Things sway in response to waves on the surface. Pockets of cold water appear mysteriously, dragged up from the deep by eddies.

Plant and animal life in the shallows is very different from that of the deep lake. Clams are common, burrowing in a zone that is not quite sand but not quite fibrous muck. Freshwater mussels are filter feeders too, letting the currents bring food to them. Forests of so-called weeds rooted in the bottom grow upward to just below the surface. Some of these were originally terrestrial plants that have since adapted to aquatic conditions. These are the plants that most often snag fishing lures and tickle the toes of swimmers. Some are broad leaved and tough, like kelp in the sea. Others are soft and feathery. Still others are stringy and fibrous like twine. When broken into detritus, they produce muck halfway between coarse sawdust and sludge.

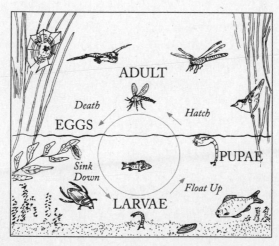

*Energy flow in lakes.*

These rooted weeds are the most important nurseries for fish fry. This is also where the majority of medium-sized fish seek food, while attempting to hide from even larger fish lurking about. Yellow perch, pike, and walleye, largemouth and smallmouth bass, and various kinds of panfish—sunfish, bluegills, pumpkinseed, crappie—are by far and away the most common fish in this zone across most of the fringe. Here the food web isn't much of a web. It's an opportunistic free-for-all based mostly on size and luck.

The top dogs are the muskies, which, depending on their mood, either cruise like sharks or lie in wait like moray eels. They will prey upon anything they think is alive and are especially fond of taking waterfowl from below. This is something I learned as a child when I discovered what I thought was a yellow bathtub duckling in my grandfather's tackle box. To my surprise, it was a fishing lure designed for muskie, equipped with a treble hook about two inches across and a wire leader so heavy one could call it a cable. Muskies don't chew their food; they inhale it whole, and have back-pointing needle-sharp teeth to make sure that nothing can escape. A colleague once told me of a fisheries biologist who dangled his feet in the water to cool off, only to have them mauled by one. Muskies tend to be absent in small kettles, where the apex predator is usually a close cousin, the northern pike, an ambush predator that patiently hides in the weeds like a submerged log until it strikes with lightning speed and ferocity. The smallest members of the pike family are pickerel. Thoreau and John Muir were especially intrigued by their mixture of exquisite beauty and savagery.

Many plants occupy the netherworld between land and lake. Lily pads grow in somewhat protected sites, preferring water of moderate depth. Their parasol-shaped leaves create deep and crisp shadows separated by shafts of light, perfect cover for camouflaged predators. Closer to shore, plants boldly emerge from the water to perform photosynthesis in the air. Teal green bulrushes, dull green reeds, and waxy green cattails turn different shades of brown in the fall and remain standing through the winter. Cattails sprout from a submerged mesh of roots, rhizomes, and tubers running within the dank muck, and have flowers that look like felt-textured hot dogs on a stick. Rushes and reeds have narrow circular stems with either pointed or flowery tips. Sedges look like grass but have stems with triangular cross sections, and often grow in clumps called tussocks.

Wild rice and other aquatic grasses grow in soft-bottomed, well-protected bays and isthmuses. They are especially attractive to water-fowl, which can be seen dabbling in and dredging the muck for seed heads and aquatic invertebrates. Swimming below them are the so-called bottom-feeding, or rough, fish of the catfish family and the suckers of the carp family. Where they pass, the water is roiled brown with muck.

Some weeds, especially the most alarming invasive species, have given up roots entirely and float instead of staying in one place. In small ponds and quiet bays, a fresh bloom of duckweed can resemble a coat of fluorescent lime green paint. Larger floating plants often have foliage arranged in whorls around a single, cablelike stem, being feathery and stringy at the same time. They grow into each other, forming great knotted masses that drift across open water and pile up when blown together by the wind.

Gliding in a canoe over submerged vegetation in clear water is like flying in an airplane above the forest canopy. Paddling through duckweed leaves a wake of dark water with turbulent eddies that slowly close back in. Drifting over lily pads is like sliding across a wet tiled floor. Entering a stand of bulrushes is like passing the tip of a comb through hair. Hitting a clump of cattails results in a rubbery collision.

In the vegetated shallows are water bugs, ranging from the tiniest red mites to water beetles large enough to eat frogs. Most of the fast dark ones are technically not bugs but beetles of some sort, each with a distinctive form of locomotion. Whirligigs spin in circles on the water's surface. Water striders (aka Jesus bugs) appear to skate. Below the surface, a solitary water boatman looks like a scull being rowed underwater. Diving beetles resemble miniature loons. Water scorpions walk around on the bottom like the real thing, and there's even an aquatic spider that lives in a diving bell of trapped air and spins its webs underwater.[5]

Leeches live in the shallows too. Their reputation as nasty bloodsuckers is a biased one. As scavengers, they provide an important service for keeping lakes clean. When swimming, they are actually quite beautiful, undulating up and down in black sine waves with perfect grace and symmetry. Out of the water, however, they look like writhing puddles of licorice gelatin. On a hook they make fantastic bait, but only for those who dare to touch them.

BIRDS, REPTILES, AMPHIBIANS, and mammals are common vertebrates of the shallow-water zone. The common loon has always amazed me. Though they fly well, they do so only to move from lake to lake or to migrate south for the winter. Loons spend nearly their whole lives as nesting pairs, either floating upon or diving deep below the water's surface. They are to kettle lakes what cormorants are to the sea, specially adapted, fish-eating divers. Their calls—the quivering tremolo, the haunting wail, and the cry—are evocative enough to send chills up and down one's spine.

Turtles laze about on logs and boulders, sunning themselves between underwater forays and aboveground scavenger hunts. Of all the turtles common to kettle lakes, the snapping turtle is my favorite, not because it's beautiful, but because it is such an archaic generalist. They move about like omnivorous armored vehicles.

The common frog is arguably every child's favorite near-shore creature. They rest camouflaged, near plants and shoreline detritus, eyes bulging out of the water and legs dangling down or planted for a quick jump. They prey on whatever is smaller than they are, generally insects and softer invertebrates. In turn they are a favorite food of near-shore predators above and below the water. A good fishing technique for bass is to toss a frog on a snag-free (weedless) hook just beyond the edge of lily pads. Given the abundance of tadpoles in spring, one could be led to believe that they exist only to feed things higher up the ladder.

Beavers are common only on small, isolated kettles and deep bays. In the East they usually dam up the ponds they occupy. In the heartland, where lakes are abundant, they need not build dams at all, so they create lodges or dens within submerged banks. These ecosystem engineers are preprogrammed to react to the sound of running water and the sight of wood, their brains being smaller per unit of body weight than any other mammal's. They also have the odd habit of passing their food through their gastrointestinal tracts twice in order to extract the nutrient they need from wood and bark. Muskrats look and act like small beavers, but they are more closely related to the herbaceous voles that live unnoticed in nearby soils. The most important role voles play in the small-lake system is to create large voids in the soil that enhance infiltration and seepage, thereby raising water quality.


THE SHORELINE HABITAT is compartmentalized along the gradient of wave energy. In the erosion mode, shorelines are invariably steep, narrow, and mantled with cobbles and boulders. With little in the way of bedrock in most kettles, this provides the only place where creatures like crayfish have a chance to live in the crevices between stones or, in the case of snails, to lick algae off their surfaces. Any sand that was originally present on such boulder-cobble-pebble beaches has been winnowed away, either dragged offshore to produce a rippled sheet or pushed down-current to a more protected spot.

Where there is a modicum of wave energy, the shorelines are usually marked by narrow beaches—ribbons of wave-washed sand and gravel extending only a few feet on either side of the shoreline. There is an almost perfect correlation between the amount of wave energy available and the texture of the sediment, with pebbly beaches characteristic of sites just barely able to hold on to the sediment and fine-sand beaches characteristic of those with just enough wave energy to keep rooted plants like horsetails from growing. It is just above these sand shorelines that the effects of ice movements are most dramatic. Masses of sand frozen into the ice are thrust ashore into piles later gouged by floes.

At the waterline is a flotsam of twigs, reed fragments, seeds, leaves, and the disarticulated remains of insects and crayfish. Sometimes there's a dead minnow, fish, or bird that washed ashore. These nutritious prizes are scavenged by nighttime opportunists such as skunks, feral cats, raccoons, coyotes, or opossums. Below the water surface are often schools of minnows whose dark shadows are easily seen against the backdrop of light-colored sand. Crayfish scavenge this submerged sandy netherworld for detritus of animal origin. They walk as boldly below the shorelines of kettles as do their lobster cousins on the floor of the sea.

Wetlands form where there is sufficient low-lying land and where the physical energy is low.[6] Fine-grained mineral sediment suspended anywhere within the lake settles best in such settings. But the majority of wetland sediment consists of dead organic matter. Peatlands, also called mires, have thick organic soils that usually devel-

*A kettle marsh in northern Michigan.*

oped as the water table rose.[7] They are boggy when the nutrient is limited and the soils are acid. This is the default condition for small kettles because well-drained, sandy soils surrounding the depressions tend to be infertile, and because bog vegetation prefers the cold microclimates of isolated basins. Bogs often have floating margins, composed by the interwoven roots of peat moss, sedge, and dwarf shrubs, which grade landward into larger shrubs and trees, typically conifers like spruce and tamarack.[8] The most common bog moss, called peat moss or sphagnum, was used as disposable diapers by the Ojibwe. Patches were simply torn off the surface, allowed to dry, and padded around a swaddled infant.

Peatlands that are richer in nutrient are called fens. Typically they support more luxuriant herbaceous vegetation. America's most well-known fen gave its name to Boston's Fenway Park. Swamps may or may not be underlain by thick peat. Regardless, they are forested wetlands, usually dominated by red maple, alder, and cedar. Marshes are herbaceous wetlands composed of various grasses, reeds, and rushes, and bordered by thickets of brush, especially willow.

Many small kettles that were once lakes and ponds are now swamps and marshes. They were either shallow enough to have already filled with sediment or rich enough to have produced a large surplus of organic matter year after year. Such infilling is the inevitable fate of all lakes created by some past event like glaciation. Most kettles have many millennia remaining before they will be covered over with wetland vegetation.

INLAND FROM THE SHORELINE is the freely drained soil of the terrestrial realm, home of *Homo sapiens*. This habitat, which constitutes anywhere from about 10 percent to more than 90 percent of the local area, is always underlain by some kind of mineral sediment, usually rock, gravel, sand, silt, or clay; otherwise it would decompose downward to the water level because a surplus of oxygen is available. The most common kettle-edge habitat is a forested bluff stabilized at a straight, steep angle after meltdown collapse, usually twenty-five degrees or more. Another common edge configuration is a sand plain or fan of sediment that washed into the kettle from the side after the basin had formed. This realm can be as treeless as the moorlands of the foggy Atlantic islands, the prairies of dry western lands, or a parking lot for a mall. More commonly, the default vegetation toward the north is pine forest, mixed with birch, spruce, fir, and sugar maple. Toward the south, the pine is mixed with oak, ironwood, ash, and a variety of other deciduous trees. Kettles within urban areas are surrounded by a continuum with pavement and buildings on the one end and houses on large landscaped lots on the other. Kettles in accessible rural settings are usually developed for recreation, with dwellings, lawns, and landscape plantings that quickly give way to natural vegetation farther inland.

Mammals and birds garner the most attention. The largest and most impressive grazer is the moose, which is generally restricted to the northern part of the glaciated fringe. Whitetail deer are more common to the south and east, nearer the deciduous forests, especially acorn-producing oaks. Elk, red deer, and even buffalo were formerly prevalent along the southern edge of the glaciated fringe. Contrary to popular opinion, whitetail deer became important in the north woods only after farming began, though they have always been important to the south, where mast forest is present. Ground-nesting birds like wild turkeys and partridges are widespread in the south and north, respectively. At the top of the local terrestrial food web in the fringe are mammalian carnivores. Minks, weasels, and ferrets go after small prey. Foxes, bobcats, coyotes, and wolverines take the middle-sized stuff. Wolves and cougars went after large prey until they became locally extinct, except in the wildest settings. Raccoons, opossums, and black bears are the most familiar omnivores. Badgers are specialists of the edge between prairie and forest, which is why Wisconsin is known as the Badger State. Beneath the surface are all manner of creatures, burrowing, digging, and living within the soil.

. . .

THE HIGHEST HABITAT OF THE kettle lake system is the atmosphere: Great blue herons fly in a straight line like jets from one marsh to another to feed on frogs, small fish, insects, snails, and worms. Bald eagles soar in circles, either cruising the lake perimeter like the fish carnivores below them or in tighter circles on air currents called thermals. Periodically they swoop down to pluck an unsuspecting fish with their talons. Flocks of Canadian geese use the lake primarily to rest, to avoid predators, and as an open latrine. Being grazers, they typically feed elsewhere, in the process transferring nutrients to the lake system, enriching them for better or worse. A variety of ducks are also common in season.

Most of the aerial action is less magnificent. Moths dominate the nighttime skies. Dragonflies are amazing aerialists, resting on reeds, the tips of fishing poles, and anything solid between their carnivorous forays. Female mosquitoes, as adults, torture mammals and birds in search of a blood meal. As juveniles, however, their air-breathing larvae are an important part of the food web. They float just below the surface of quiet water, gorging on plankton and being fed upon by small fish. Biting insects like horseflies can tear off chunks of an animal's skin. Almost as annoying are swarms of tiny flies called no-see-ums. Feeding on all kinds of flying creatures are gnat-catching swallows, moth-catching bats, and bird-eating hawks. Scores of mid-sized songbirds from chickadees to robins come and go in season.

THE LAKE-FOREST ECOSYSTEM is more complex than the sum of its parts, making it an emergent system. Within it, life flows in all directions: from the bacteria of the deepest muck to the highest soaring eagle; from the center of open water to the hills of the inland forest; and from the quietest marsh to the noisiest, wave-battered boulder peninsula. The molecules from which every creature is made cycle perpetually between air, water, soil, and sediment according to fundamental laws set by the physics and chemistry of natural waters. Given its many checks and balances, this system had been fairly stable for approximately ten millennia. But it had evolved no protection against the biggest thing to hit kettle lake country since meltdown— the arrival of human beings from Europe with a cultural template for permanent residency on private property, and the power to reshape the world.

## 10.

# Loving Lakes Too Much

SOME FORM OF HUMAN ACTIVITY has impacted every kettle lake and pond within the glaciated fringe. The most pristine have hidden problems associated with global air pollution. Those with farms in their watersheds, small cities on their shores, and concentrations of second homes and resorts exhibit a range of impacts. In a few cases, what were formerly liquid blue crystals in the forest wilderness have since become weedy sludge pools contaminated by toxic metals and pathogenic bacteria. They've been loved nearly to death.

My personal understanding of the human transformation of small lakes began along the edge of the prairie at Union Lake, Minnesota, where I summered as a child during the 1950s and 1960s. I recall that in the later years, the water seemed a bit less clear and the chore of raking weeds near the beach more demanding, and the sandy shallows felt squishier to my toes. As the summers rolled by, I remember catching more perch, my first bullhead, and seeing my first infestation of snails. These observations signal a shift from a leaner and cleaner lake to a murkier, more biologically active aquatic regime. Completely invisible to me were the rising counts of fecal bacteria and the increasing levels of persistent organic toxins, like DDT, and heavy metals, such as lead. At the time, I didn't make the connection between changes within the water and the rising number of cottages on the shore.

My hometown of Bemidji, Minnesota, in the north-central part of the state, has the distinction of being the northernmost point on the Mississippi River. The town's history is connected to the story of kettle lakes in almost every way: beginning as a trading post in the late nineteenth century; being named after Chief Bay-me-ge-maug; booming as a logging town in the early twentieth century; and then stabilizing as a regional commercial and government center for the summer tourist industry and the four Indian reservations surrounding it in every direction.[1] Lake Bemidji is a merged pair of large kettles with the shape of a footprint (alleged to be that of Paul Bunyan).

During the late 1960s, I was a student at Bemidji High. Our mascot was the lumberjack. During summers, my crowd hung out at the lake. Sometimes we noticed that it changed color from blue to greenish gray during what we later learned were algal blooms. We also developed swimmer's itch, caused by an invertebrate parasite common to lakes with too much nutrient. During winters I had a fishing shanty on the ice, from which I watched my neighbors dump tainted kerosene. Behind every shanty was a large mass of yellow snow, tinted by urine. This era was just before the passage of the National Environmental Policy Act of 1970, which gave the federal government the right to step in and help raise the water quality of lakes nationwide.

At the time, this small city of about ten thousand was treating its sewage by screening and settling the solids and letting the liquid trickle through crushed rocks before draining into the lake at the inflow of the Mississippi River.[2] The massive paper pulp plant on the south end of the lake used the lake as a chemical drain. This was also the era of cheap leaded gasoline, low gas mileage, and exhaust systems without catalytic converters. Hordes of summer visitors and seasonal residents caused the place to nearly triple in population each summer. The fumes from their automobiles wafted right to the lake. In late winter, melting snowbanks darkened as the soot from auto exhaust and heating oil became concentrated. In spring there was a three-week-long flush of runoff from streets, driveways, and parking lots directly into the lake. All year long a steady stream of truck traffic moved northward along the edge of the lake on U.S. Highway 2 between the grain elevators of the Dakotas and the grain ships waiting in the port of Duluth. Their exhaust was spewed just upwind of the western edge of the lake. In spite of all this contamination, life seemed idyllic. Little did we know how hard we were hitting the lake ecosystem.

The Clean Water Act was passed in 1972, midway through my years at the local state college. Our mascot was the beaver. While being introduced to geology, plant taxonomy, zoology, conservation, hydrology, and limnology in my courses, I was also a successful candidate for student government. One unrealistic plank in my environmental platform was to clean up the lake. Though things have improved in the last few decades, the headwaters of the "Father of Waters" were still running murky green in 2007, as seen on national television following the collapse of the Interstate 35W bridge between Minneapolis and Saint Paul.[3] Locating the bodies and submerged cars was stymied by the lack of transparency.

In search of a cleaner and more secluded lake, my parents bought a cabin on Lake Plantagenet, one of thousands located south of town nearer the Mississippi headwaters. The Ojibwe called it Kubbaqkunna or Rahbahkanna, which means "rest in the path." For millennia, they had stopped to rest on its sandy shorelines after paddling their birch-bark canoes through the winding marshes to the north and south. In 1832 Henry Rowe Schoolcraft and his party canoed directly below the bluff on which our family cabin is perched. Three members of his party commented in their published journals on the lake's beauty. The natural scientist Louis Houghton recalled a lake "of beautifully pure water . . . the shores of which were sandy and supported a dense growth of pine."[4] Military commander Lieutenant James Allen recalled a "pretty little lake." Schoolcraft wrote of its "pleasing aspect." On my first trip to the lake in 1970, I remember thinking that it was indeed beautiful, cleaner than Lake Bemidji, though not quite as clear as the Union Lake of my boyhood.

Almost four decades later, in July 2006, the state regulators were at the lake in full force. One officer was stationed at the public boat launch, inspecting propellers and trailers for invasive species hitchhiking between lakes. Another was checking fishing licenses, ensuring that boats were properly equipped, and inspecting the catch limit. A third official was sitting in the truck with a clipboard, filling out what I presume was environmental-monitoring paperwork. As my brother and I motored across the lake in a boat equipped with a car-sized engine, we cruised over what resembled an extremely dilute green pea soup straddling the edge of transparency. The color came from suspended algae, venting a smell like that of a hay field cut a few days earlier. Later that day I went swimming. The muck just beyond the edge of the dock felt like stiff pudding to my feet. The submerged thicket of weeds I swam through was slimy with a gelatinous film. Underwater visibility resembled the hour before dark on a gloomy winter day.[5]

Normally the lake is in better shape, especially in June and September when the water surface is colder and photosynthesis is less strong. Good years came when the water level was high because the groundwater springs were more powerful due to greater infiltration. Bad years came when the water level was low and the volume more stagnant. Plantagenet still provides nice views, especially during sunrise and sunset, the lake breezes are still nice, boating and fishing remain popular, and the regulators report slight improvement. But it doesn't come close to being the "beautifully pure water" seen nearly two centuries ago, before logging and agricultural development in the watershed and development of the shoreline for second homes.

By the mid-1980s my wife and I were living in Mansfield, Connecticut, which has so many kettles in one corner that the town was originally named Ponde Place when settled in the late seventeenth century. As a geology professor setting down New England roots, one of my first day trips was a pilgrimage to Walden Pond, for which I had developed a strong personal attachment. Because of its international fame, the degradation of Walden Pond has been unusually well documented. In the 1840s Nathaniel Hawthorne remarked, "Walden Pond was clear and beautiful, as usual . . . If I were to be baptized, it should be in this pond; but then one would not wish to pollute it by washing off his sins into it."[6] A few years after Thoreau ended his experiment in deliberate living, the Boston-Fitchburg railroad was bringing tens of thousands of visitors to the pond. Gardens were fertilized with manure. Dogs did their thing. Based on a study of aquatic microfossils from Walden Pond, the limnologist Marjorie Winkler documented a dramatic rise in nutrient pollution beginning at about the time of Thoreau's sojourn.[7] The Walden sediment archive also shows an erratic increase in charcoal caused by fires and in the amount of mineral sediment being washed in by erosion associated with logging, the railroad, and trampled trails.

The pond flora changed when the seeds and shoots of invasive weeds arrived furtively via horse dung, boating and fishing equipment, and natural dispersal mechanisms. The fauna changed when new species of minnows, worms, insects, and leeches escaped from minnow buckets and bait boxes or were deliberately introduced. By 1905 the pickerels that had inspired Thoreau to wax so eloquently about nature's crystalline beauty were locally extinct. By 1921 things were so serious that a Walden Pond Protective Association was created. During the hot summer of 1935, more than 483,000 people visited the pond to cool off, 35,000 on one day alone.[8] Early limnologists, including Edward Deevey Jr., took notice and began to investigate beyond where Thoreau left off. Alarmed by the crowds, and by the rising concentrations of pollution, scientists issued a 1939 report with the following carefully nuanced sentence: "The principal anthropogenic perturbation causing eutrophication is the hypothesized swimmer input of nutrients."[9] Translation: What had once been a pond as pure as a Pierian spring was becoming murky with algae because thousands of people were peeing into it surreptitiously.

In 1958 the residents of Concord voted overwhelmingly to build a new dump in the permeable gravel adjacent to Walden, the leachate from which will probably leak forever. In 1968 the entire water volume was deliberately poisoned with rotenone, so that "weedy" fish could be eliminated and other species restocked. This was the same year in which the National Environmental Policy Act was being drafted in the halls of Congress. But even after wiping Walden's zoological slate clean and starting over, the so-called better fish remain inedible, thanks to toxic contamination. Though Walden is clearer than most Massachusetts lakes, the input of tainted dust from the nearby urban population and urine from its many swimmers remain threats.[10]

BEING ERASED BY EROSION is the worst thing that can happen to a kettle pond. Gravel mining obliterates them through removal of the landform. On Cape Cod, the sandy cliffs behind its world-famous beaches are cutting back into kettle moraines at rates of several feet per year. In an extreme example, Billingsgate Island off Wellfleet, Massachusetts, was a Yankee-era fishing village with a lighthouse and schoolhouse, thirty homes, and dockside life. Today it's a submerged shoal exposed at low tide.[11] Rapid coastal erosion of bluffs containing kettles also takes place on Long Island and Fisher's Island, New York, in the Charlestown section of Rhode Island, and on the Elizabeth Islands off southern Massachusetts. Wasque Point, on the southeast side of Martha's Vineyard, moved back 350 feet in a single year, an order of magnitude faster than the average rate. The accelerated rise of sea level expected as a consequence of climate warming will only increase the rate of kettle moraine erosion.

Kettles also disappear by being filled. Lakes are destined to fill at a geological time scale because they are closed topographic basins, with no way for sediment to escape. The natural succession is from lake to pond to pool to wetland to mossy forest. Humans have accidentally accelerated this natural succession by increasing watershed erosion and by stimulating the growth of aquatic plants, which increases the amount of muck, which decreases the volume of the basin. However, many lakes and ponds have been filled on purpose. Before wetlands regulation, prairie potholes and other small, shallow kettles were routinely filled to create arrow-straight roads and unblemished fields on large grain farms. During the 1930 dust bowl years, the main conservation problem in America was soil erosion on hillsides. Much of what was lost from the hillsides washed into shallow rural kettles, either shrinking their size or filling them completely. Today a more important source of sediment pollution to small lakes is the sand being spread for winter road traction. In a case of double destruction, mining the sand in one place and spreading it on another destroys kettles in both places.

Closer to cities, hundreds of small kettles were filled and their adjacent kames flattened because they stood in the way of some improvement, perhaps an exit ramp, shopping mall, or cineplex. Still others disappeared when they became the sites of landfills, usually referred to as dumps before the 1970s. Kettle holes and swamps were then seen as ideal sites because they were accessible low spots that were visually shielded by bluffs. The ubiquity of sand and gravel also kept construction and maintenance costs low. Unfortunately, these materials were highly permeable. Rain and snowmelt, having been steeped with garbage within the cup-shaped silt liner for decades, now overflow as toxic tea.

Throwing or dropping stuff overboard contributes insignificantly to anthropogenic filling of lakes. Instead, the problem it creates is chemical. Megatons of lead shot have been sprayed from the barrels of shotguns by duck hunters. For more than a century, lead sinkers for fishing have been lost overboard or off the end of a dock. More unusual metals have leached into the water from lost anchors, fishing tackle, poles, and minnow buckets. Corrodible beer and soda cans were pitched overboard without concern before the era of recycling. During ice carnivals and spring "ice-out" lotteries (wagers on what date ice will leave a lake or river), junk cars were hauled out onto the ice and left to sink, their fluids free to disperse, their asbestos-lined brakes free to disintegrate, and the copper in their alternators and lead in their batteries free to dissolve and disperse. In Salem, Connecticut, there's even a lake with a piano resting on the bottom. It fell through the ice when someone was trying to move it to the opposite shore.[12] Lakes were ideal stages for community fireworks displays, leaving chemically tainted cartridges to sink and dissolve. The original two-cycle outboard motors required that motor oil be mixed with gasoline, some of which would spill into the boats and be bailed out into the lake after the next rain. Plastics are still an issue, especially durable monofilament line. It can physically trap creatures when in a tangle, restrict their function when wrapped around a limb or gill, or be fatal when ingested.

Kettles are also being adversely transformed by human efforts that physically change the shoreline. Most dramatic is when lakes are artificially raised with a dam. Sometimes this has the effect of converting several coalesced small lakes into a single larger one, inaugurating a cascade of ecological change. In cases where kettles are fed by small streams, raising a dam will increase water loss by seepage and evaporation, which may reduce the flow to downstream systems.

Most shoreline alteration is done on private property to enhance shoreline access for recreational purposes. It's one thing to remove a patch of trees, build a cute cottage on concrete blocks, and set out a dock over a small fringing wetland and sand beach. It's quite another to clear-cut two acres, build a multibathroom McMansion, roll out a chemically treated lawn, excavate the shoreline, and replace it with a wall of boulders to prevent wave erosion and cut back on the mosquito population. Doing so may be legal in some places, but it destroys the natural habitat of the lakes for wildlife.

The beautiful bay on Union Lake that I used to fish in as a child—with its concentric canopy of trees, clutches of cattails, stands of rush, and sheets of lily pads—is now mostly a boulder-armored shore. Human sounds are amplified. Natural ones are quieted. In the 1960s dozens, if not hundreds, of frogs jumped out of my way every time I walked the shore, and thousands of minnows could be seen darting about in schools. In July 2008 I didn't see a single frog or minnow when walking the same path, and a guy I talked to who was repairing a cottage porch complained that he hadn't had a nibble in three nights' fishing. Sheets of black plastic, now torn, had been laid down on the sand, presumably to prevent weed growth. The biologically rich lake of my memories is still scenic and cool to the touch on hot days, but it's eerily quiet, except for adult voices, boat noise, and dog barks. In a place that used to be crawling with children for weeks on end, I saw only one pair of adolescents standing quietly on the opposite shore.

Even on lakes that aren't fully developed there are usually a few uncooperative types who do their own thing, regardless of the rules. Over the last few years I've kept my eye on a lake lot north of my parents' cabin on Plantagenet. Local zoning regulations logically prevent the construction of roads leading down to the shore because they compromise the view for others, lead to overbuilt private marinas, and increase the sediment flux. So what did the owners of this lot do? They built a wide, evenly graded "trail," accessing the beach via car-sized all-terrain vehicles equipped with tractor tires. Following rainstorms, the sand eroded from the gash in the bluff washes down to the shore, forming a delta. In 2006 I wrote an official letter of complaint. But when I boated by this lot in 2008, a large yellow bulldozer had just left its cleat marks on the trail and was reworking the delta into "improved" shoreline. This travesty may comply with the letter of the law, but certainly not its spirit. And if what they did is indeed against the law, I suspect a small fine was paid and that was the end of it. Community pressure by the local property owners association seems to have made no difference.

Lake pollution is mostly invisible. Global chemical junk traveling high in the atmosphere—radioactive fallout, heavy metals, and persistent organic pollutants (POPs)—reaches even isolated kettles as aerosols—tiny particles of dust, soot, and droplets that either fall directly on the lake surface or are washed in by runoff or groundwater.[13] Most POPs are agricultural insecticides that were sprayed or dusted on crops before the world realized how long they would last, though many are still allowed. Unfortunately, they bioaccumulate in the food chain, their toxicity becoming more concentrated with each step. The best-known, dichloro-diphenyl-trichloroethane (DDT), is now banned. Today the worst offenders are various polychlorinated biphenyls (PCBs). Other nasty-sounding POPs include dioxin, chlordane, dieldrin, aldrin, toxaphene, mirex, hexachlorobenzene, endrin, and heptachlor. All are raining or leaching into a kettle lake near you.[14] Other pesticides, Atrazine for example, are not POPs but are damaging as well, especially to amphibians.

The story with heavy metals is more complex. At low doses, natural elements such as selenium, cadmium, and copper are vital for metabolic processes. At high doses, they are dangerous, even fatal. Metals reach lakes through direct atmospheric deposition, stormflows, contaminated groundwater plumes, and from objects within the water. Each metal has its own sources, aquatic pathways, and medical consequences. Arsenic is in the lake news today because the change in chemistry from oxygen-rich to oxygen-poor conditions has increased the presence of this carcinogen. Lead and mercury are well-known neurotoxins. Lead dust from the gasoline additives used decades ago is still washing into lakes, along with lead leaching from old paint chips, corroding metal pipes, and other sources.

Mercury usually arrives from more distant sources, typically coal-burning power plants, trash-to-energy facilities, and factory smokestacks operating just below the legal limit. The main concern about mercury involves its transformation from elemental mercury into a deadly neurotoxin called methylmercury, which bioaccumulates in tissue. Methylmercury is more effectively destroyed by natural light when a lake is clear, giving us yet another good reason to keep murkiness under control. On outer Cape Cod, far from the nearest smokestacks and factories, the lead concentration in some ponds has risen twentyfold during the historic era, the mercury concentration sevenfold. Diatoms are being born with deformities.[15]

A growing concern is water pollution caused by pharmaceutical waste associated with humans and livestock. Much of the population of the glaciated fringe routinely takes prescription medicines, notably antibiotics, steroids, and synthetic hormones. Many are excreted from the body with some or all of their potency still intact, then flushed into the wastewater stream, where they enter streams, aquifers, and then lakes. Particularly nettlesome are, antidepressants, which interfere with the reproduction and development of susceptible organisms; antimicrobials, which help drive mutations toward drug-resistant forms; anticancer drugs, which can be toxic; and anticonvulsants, which interfere with motor activity.

Some unlucky kettles are being poisoned by the very groundwater that feeds them. The most famous case involves those near Otis Air Force Base on Cape Cod. For military purposes, it was completely logical to site a large airfield above the widest part of the sand plain that forms the inner cape. For environmental purposes, one could not have picked a worse place, directly above the crest of a groundwater mound draining outward in all directions through highly permeable materials. Prior to the 1980s those responsible for waste disposal deliberately pumped millions of gallons of solvents, fuels, soaps, and special fluids into the ground or flushed them down the drains, expecting them to disappear. Ever since then, they've been seeping outward in multiple plumes from the concentrated center in a pattern shaped like the petals of a daisy. The longest one, now over five miles long, passes right into and out of Ashumet Pond in Mashpee, an otherwise beautiful kettle lake, now a poster child for environmental callousness.[16]

Little noticed three decades ago were invasive and introduced species. Now they rank in many states as the single most pressing problem degrading aquatic ecosystem function and reducing property values. Eurasian milfoil and fanwort are choking fertile lakes with floating mats of weeds sometimes too thick to boat through. Water chestnut and hydrilla can blanket the whole surface, suffocating the bottom. Pondweed fills the upper few feet of the shallow water like a mass of cotton candy. Purple loosestrife, sold as a garden perennial, is on the loose and changing wetland function for the worse. Phragmites, a common reed taller than a field of prizewinning hybrid corn, is shading out native reeds, which are lower and grow less vigorously. Zebra and quagga mussels cover and clog boat parts and water intake or outflow infrastructure as do barnacles in the sea. The rusty crayfish is a bully, taking over this scavenging niche with unknown consequences. Escaped minnows are changing food webs. The reintroduction of beaver is submerging wetlands and changing nutrient loads in unknown ways.

Stocking carnivorous game fish involves the deliberate introduction of a species, rather than a natural invasion. This can lead to fewer medium-sized predators, which can mean more zooplankton-eating fish, which can mean fewer zooplankton, which can mean more algae. The opposite can happen as well, depending on the details. The antithesis of stocking is overfishing, often encouraged by fishing derbies. This can be either beneficial or destructive, depending on the circumstances. Trying to artificially simulate the checks and balances within complex food webs is arguably impossible. A wait-and-see management plan is the usual result.

Public health issues caused by pathogens and parasites also pollute lakes. The number one issue at many public swimming beaches is an unexpected outbreak of fecal coliform bacteria, which can lead to closures of beaches during the dog days of summer, when they are most needed. Outbreaks usually originate after animal and human wastes flow into a lake as runoff or leach from shallow groundwater plumes. Other waterborne diseases formerly restricted to more southerly climes have now reached the northern glaciated fringe. Malaria, West Nile virus, and eastern equine encephalitis are mosquito-borne diseases on the rise and correlate with the warming trend of the last half-century. Disease-carrying invertebrate aquatic pests are also increasing.

Acid rain is another invisible problem for many kettle lakes. The acidity comes from fossil fuel smokestack and tailpipe emissions. Sulfur and nitrogen combine with atmospheric water droplets to create dilute sulfuric and nitric acids, which brings the pH down to well below that of the "natural" acid rain. Higher acidity enhances soil weathering, which changes the chemical balance of the water reaching the shoreline by seepage. A low pH, for example, might restrict the precipitation of shell carbonate by clams and snails, or eliminate an important larval stage in some critical part of a life cycle. Acid rain is naturally buffered by lakes with limestone watersheds. Most kettles, however, occur in silica-rich materials that lack such buffering capacity. Artificial buffering is done by sprinkling crushed limestone on the surface.

A classic kettle on Cape Cod illustrates how tightly linked parts of the system are chemically linked.[17] At Gull Pond during the decade from 1982 to 1991, the average annual rainfall decreased from about sixty to thirty inches. Accompanying this change, but lagging two years behind it, was a dramatic increase in the acidity of the pond, its pH changing from slightly to moderately acidic. This wasn't caused by a change in the arriving rainfall, but by a change in the groundwater flow. Less rain caused less infiltration through the sandy soils of the pine-oak forest, causing less calcium, magnesium, iron, and sodium to be leached into the water to neutralize the rainfall. Meanwhile, the reduced precipitation caused a small drop in the water table within wetland organic soils. This oxygenated the peat, converting sulfur into soluble compounds that increased the acidity. The community of algae shifted, with an ecological ripple effect propagating upward through the food web to the human beings who went there to catch fish.

THE THORNIEST PROBLEM FOR KETTLE LAKES is being overfed with nutrient.[18] At low or moderate concentrations, nutrient is essential for lake life. But beyond a threshold unique to every lake, it's considered a pollutant. Usually, it's the phosphorus that's the problem. The worst outcome is when a previously transparent rural lake of crystalline beauty develops the greenish gray, sludgy look of an urban duck pond. Almost as bad is when clear water is covered with green scum or a carpet of filamentous algae so thick that birds appear to walk on water. When large weeds are the problem, entire bays can become miniature Sargasso Seas nearly covered with floating rafts. Clots of blue-green algae (cyanobacteria) resembling torn fragments of felt—gray on one side and black on the other—can float up from a slime-coated bottom. Blizzards of lake flies can fall so thickly after their short lives expire that their remains have to be shoveled and plowed off streets and driveways.

Decomposing the excess growth of aquatic plants in an overfed lake usually requires more oxygen than is available. For fish such as trout that require cool-water conditions, a decline in oxygen forces them up into warmer water, where they may not survive, causing fish kills.[19] Anoxic conditions leading to fish kills can also happen during the winter beneath the ice. A mass of water with no oxygen at the bottom is referred to as a "dead zone."

When the bottom bacteria that depend on oxygen for respiration can't survive, those that depend on fermentation, sulfate reduction, or methane-producing pathways take over. This can lead to outbreaks of botulism or salmonella, which kill fish and waterfowl, which then drift to shore and rot. Vented from some algae under certain conditions are toxins called microcystins, which, when breathed by humans, can impair health. The rotten-egg odor of hydrogen sulfide, the carrion smell of worms, and the septic tank smell of everything else can waft up from the muck. Methane can bubble upward through the water column, reacting spontaneously with whatever dissolved oxygen is left at higher levels and killing tiny animals near the surface. If the methane escapes the water, it will spontaneously oxidize or even combust, producing an overfed or underoxygenated lake's most magical lake effect, a "will-o'-the-wisp." At night these yellowish orange splotches of flame look like ghosts swirling over the surface.[20]

The most common problem associated with slightly overfed lakes is olfactory. The piney crispness of lake air is gone, replaced by the halitosis of heavy metabolic breathing within the water column, especially at depth near the muck. The most common gases vented from lakes don't have a detectable odor when pure, but are easily perfumed with more fishy-smelling, volatile gases created by low-oxygen conditions. Fermentation exudes a vaguely breadlike odor. The sulfurous smell of salt marsh is objectionable when it seems so out of place. A change in the breeze can often be smelled before it is felt.

The optical properties of the water also change with nutrient excess. Increased algal growth causes light to penetrate less deeply and change color, scattering back a milky green instead of a clear blue. One scientific measure of this color change is called the chlorophyll-a index, which is widely used as a measure of biological productivity.

Sometimes a lake in late summer is clear at the surface but full of algae at depth, where cooler temperatures or more nutrients prevail.[21] At such times, a swimmer can dive down through clear water only to find it murky at depth. The deep turbulence caused by propellers and jet skis can roil phytoplankton up from below that resemble the contrails of jet aircraft, but milky green in color. At other times, especially following a soaking rain, a clear lake surface can flash green as the algae rise up to capture the fresh influx of nutrient.

Lake-by-lake variation in nutrient content is completely natural. Flat topography produces lower nutrient loads because there is less runoff and therefore less chance for erosion to suspend and transport particles with adsorbed phosphate. Watersheds draining through marine sedimentary rocks produce more nutrient than those draining through granite and quartzite. Coarse-textured materials like gravel allow nutrient to infiltrate quickly toward a lake, whereas clay and humus soils help bind it in place. Soils close to neutral pH are more susceptible than those that are either significantly acidic or alkaline. Any natural change in the watershed from, for example, a forest fire, a hurricane blow-down, a drought, or an insect invasion can tip the nutrient balance in the positive direction for a decade or more.

Nutrient pollution usually comes from invisible sources distant from the lake. It comes with the wind as dust, soot, tiny droplets, and so forth. This is a particular problem for lakes downwind from cities, major interstate highways, and factories or plants that burn fossil fuels or incinerate waste. Logging, even for firewood, can cause a pulse of nutrient pollution released by decomposition of stumpage and woody delons and washed to bodies of water by enhanced runoff.[22] Any molecules of phosphorus, nitrogen, or carbon not reclaimed by fresh terrestrial growth will likely wash downstream and enhance the growth of aquatic plants. The conversion of pasture to tillage also increases the nutrient flux because the latter is usually more heavily fertilized.

The application of manure or chemical fertilizers to the terrestrial soils of farms, gardens, golf courses, parks, and residential lawns is done for the sole purpose of stimulating plant productivity. It follows axiomatically that any fertilizer not taken up by terrestrial plants or absorbed in the soil will eventually reach the lake, where it will do the same thing, but with aquatic plants. The tight correlation between the timing of agricultural fertilization and the timing of algal outbreaks is routinely confirmed by chemical spikes in the water during or shortly after the application. Nutrient loads from cultivated fields can be up to ten thousand times higher than under forested conditions.

The U.S. farm population peaked in 1916, just about the time the seasonal lake population began to rise across the glaciated fringe. During the first few decades of recreational use, the water within lakes typically remained fairly clear, even as their shorelines were developed with second homes and resorts, because the nutrient added by human waste, detergent, and garden and lawn fertilizers was being absorbed by unseen increases in the growth of shoreline plants. But beginning in about 1950, farm productivity began an exponential increase directly related to the heavy application of manufactured fertilizers. By the time usage peaked in 1981, when 54 million tons were applied, the buffering capacity of many lakes to absorb nutrient had been exceeded.

The average human excretes several grams of phosphorus per day. Some of the phosphorus pooped into outhouses generations ago is still working its way to the shoreline. Much more arrives from improperly designed or malfunctioning septic systems, which consist of a tank where the solids settle and a drain field where the liquid is supposed to be cleansed by bacterial action as it soaks into the soil. Malfunctions of these systems are hard to detect because the whole process is underground and because malfunctions often occur during heavy rainstorms, when nobody is watching. This is why pollution from on-site wastewater systems is considered nonpoint. Larger wastewater systems from resorts and small municipalities are more carefully regulated as point sources. Regardless, every molecule makes a difference.

Bodily wastes from pets can add to the problem. Dog chow is

*An outhouse near Lake Plantagenet, Minnesota.*

made from a mixture of vegetable products and the meat of animals that were fed with grain that was fertilized by phosphorus. When dogs defecate on the ground, the molecules of phosphorus redistribute within the soil and transfer to the lake by runoff or groundwater infiltration. Domestic cats are less a problem because their litter is usually tossed in the trash and hauled away to a landfill.

Other animals besides humans and their pets void bodily wastes that contribute to the problem. On dairy farms where the cattle graze healthy pastures, most of the nutrient dropped as manure is recaptured by plants before it reaches the nearest lake. Not so with livestock feedlots and equestrian facilities, where the nutrient is imported as hay and grain, and where excrement, rainfall, and heavy trampling create a concentrated nutrient porridge that almost always leaks downstream. Sometimes the biggest problem for a recreational lake is something like a poultry farm many miles away. The feeding of geese, ducks, and gulls near lakes is also a problem because these waterfowl excrete nutrient that finds its way to liquid water. The same thing is true with the feeding of deer. Artificially high populations lead to the overgrazing of nutrient-rich shrubs and their conversion into water-soluble feces.

The soap and detergent industry dumped a huge amount of phosphorus into the water system during the 1960s and 1970s, some of which is still causing problems.[23] Cleaning agents were manufactured with a heavy dose of highly soluble sodium phosphate because it helped remove organic residues from laundry, dishes, and human bodies. In a case reminiscent of the tobacco industry and the climate change debate, a few lake scientists in the pocket of the detergent industry argued that phosphorus wasn't a problem at all, claiming that nitrogen—or in some cases carbon—was the main culprit. Though the battle over phosphorus was finally won during the mid-1970s, it took more than a decade for this problem to be phased out of retail products. But the legacy of that industry's damage to lake ecosystems lives on.[24]

Halting the flux of nutrient to a damaged lake from external sources often does not stop the problem. Above some threshold unique to each lake, an excess accumulation of organic matter pushes the demand for oxygen in bottom waters beyond a tipping point. Metabolic pathways change as "good" bacteria give way to "bad" bacteria. What was previously a sink for external phosphorus in an aerated condition becomes an internal source that will last until the excesses of the past are burned away, or until the lake bottom becomes oxygenated again. (A drop in the water table will also convert marginal wetlands into sediment sources rather than sinks.) Once a lake reaches the tipping point, there are only two reasonable management alternatives.

The first is to put the lake on a strict diet from all external sources. This is done by regulating fertilizers and bodily wastes more tightly and by capturing as much of the surface drainage as possible and forcing it to infiltrate downward to the groundwater system, rather than run directly into the lake. This can be accomplished in a variety of ways, notably with porous pavements and small catch basins called rain gardens. Minimizing runoff prevents phosphorus molecules adsorbed onto suspended particles from traveling like ticks attached to a host. If a lake can be successfully starved from the outside, it will be forced to live off its detrital fat until things stabilize again at a more appropriate level of productivity. This may take decades, but it will eventually work.

The more radical solution is to deplete the lake's internal storage of phosphorus. Enriched water just above the muck can be pumped away. Alternatively, the dissolved phosphorus can be precipitated to the bottom by adding clay, carbonate, or alum to the water. Other techniques involve fiddling with the muck rather than the water, because it's the main source of internal phosphorus in most lakes. Oxygenating the bottom water, either with bubblers or through the application of nitrate, can deactivate this source. The muck can also be vacuumed up from the bottom and trucked away as if it were the sludge from a septic tank or cesspool. Think of this as lake liposuction. Alternatively, weeds can be allowed to grow before being cropped and exported, the excess nutrient being carted away. Think of this as an underwater haircut. Large mechanical weed harvesters resembling a chimera of barge, hay cutter, and wood chipper are becoming more common.

An alternative to fiddling with the water, muck, or weeds is to fiddle with the biota and the human behavior instead. Poisoning catfish, carp, and ducks or outlawing jet skis and high-horsepowered motorboats from shallow water can reduce the stirring of the sediment and the phosphorus flux by about half. Rounding up and euthanizing resident waterfowl will help. Ecologists can tweak links in the food web in an attempt to bring down the production of organic matter or raise the level of oxygen. There's no shortage of tricks up the sleeves of experienced lake managers and engineers. The main problems are making residents and users aware that something can be done, deciding which techniques to implement, finding the money to pay for the remediation, and having the patience to wait for signs of improvement.

. . .

IN 2000 CONGRESS ASKED ITS LEAD AGENCY for lake research, the U.S. Environmental Protection Agency, about the status of American lakes. The agency's answer was effectively "We don't know."[25] This ignorance generated funding for a *National Lakes Assessment*, designed to produce a statistically robust snapshot of lake status by the year 2010, based on recreational, ecological, and chemical indicators, one that can be compared with previous assessments from the late 1980s and early 1970s. The sample will cover all fifty states, with 909 randomly selected lakes to be visited once and 91 to be revisited. The sample includes so-called reference lakes, those judged to be closest to pristine conditions.

Since the assessment wouldn't be available before the publication of this book, I tried to come up with my own, based on a cursory review of the technical lake literature, state-by-state agency Web sites, a roundtable discussion at a 2007 meeting of the North American Lake Management Society, and ad hoc interviews with eminent lake scientists.[26] One problem with such an exercise is that the quality of any small lake is hard to define, being subject to value judgments. Are they playgrounds for water sports? Reflecting ponds for private parks? Natural areas? For a seventeenth-century voyageur or a nineteenth-century timber baron, the quality of a lake was measured by the efficacy of travel or the ability to float logs, respectively. Fishermen who prefer bass to trout might argue that lake quality improves with additional nutrient. Natives who depend on their rice harvest prefer moderately rich lakes, not lean ones.

The criteria I used were nutrient pollution, invasive species, shoreline habitat destruction, disease-causing bacteria, acid rain, and chemical contamination by toxic metals and synthetic organic compounds, including pharmaceuticals. My general conclusion was that approximately one third of kettle lakes across the glaciated fringe are in good shape. Generally speaking, these out-of-the-way lakes have yet to be "improved" by development but receive increasing fallout from contaminated aerosols. The remaining lakes, most with significant recreational development, are roughly divided between those that have been significantly compromised and those whose condition ranges from poor to terrible owing to a host of problems.

At first I found it ironic that we know so much about a few individual kettles—Linsley Pond in Connecticut, Mirror Lake in New Hampshire, Walden Pond in Massachusetts, Trout Lake in Wisconsin, Kettle Lake in North Dakota—yet so little about the whole population. But this makes sense. Small, isolated, average-looking, and mostly privately owned bodies of water generate little attention beyond their locally owned shorelines.

## 11.

# Lake Futures

MODERN LAKES STAND ON THE PRECIPICE between past and future. On the horizon are three potentially harmful megatrends coming at us like freight trains. How we deal with them will determine the fate of America's kettle lakes.

The first megatrend is overdevelopment of lake shorelines. The psychic advantage of waterfront living is undeniable. Thanks to unprecedented levels of wealth and longevity, millions of retirees are now enjoying their "golden pond" years on lakes all year long. Millions of younger workers can now use portable electronic communication devices to telecommute from lakeside homes. And a well-maintained network of excellent roads leading away from cities makes what may have been a rustic weekend cottage or cabin fifty years ago a renovated year-round home today. Most kettle lakes lie within easy striking distance of Minneapolis, Milwaukee, Chicago, Detroit, Toledo, Columbus, Buffalo, New York, Hartford, Boston, and dozens of other metropolitan districts. These changes, combined with a shift toward an economy based on information, have steadily increased the demand for lakeshore property.

The desire to live on lakes year-round is especially acute for baby boomers like myself who developed a childhood attachment to lakes, even more so for those who inherited the family cottage, cabin, or camp from their parents. All too often, such inheritances are followed by the razing of the old summer place and the construction of a multibathroom house on a scale inconceivable a half-century ago. If anything, the diminished use of fossil fuels projected for the future

*Intensive development for family recreation at Elk Lake, Michigan.*

will increase the incentive for lakeshore owners to spend more days at the lake, rather than in town, sending more water laden with chemicals—nutrients, pharmaceuticals, detergents, toxic substances—down their bathroom, kitchen, laundry, and workshop drains.

CONCERNS ABOUT THE RISE OF SEA LEVEL, increased hurricane strength, and the cost of insurance premiums for seaside properties are raising the demand for freshwater shores to astonishing levels. The front cover of a magazine recently described a wooded lakeside scene somewhere in the middle of Connecticut as "the new ocean."

Novelist and essayist Jim Harrison wrote about the kettles on the thumb of Michigan's mitten north of Detroit. The woodsy rural landscape he knew as a boy has "gone in thirty years from a basically agrarian and commercial fishing enclave to an elaborate playground." There, in the "late sixties, vacant waterfront was sold for about three hundred dollars a foot. Recently a two-hundred-foot lot sold for twelve thousand a foot, bringing the total to well over two million dollars."[1] My parents bought their three-bedroom cabin and a hundred feet of bluff-backed, sand-beach Minnesota waterfront in 1970 for eight thousand dollars, a mere eighty dollars per foot. Today it would cost that much to redo the bathroom. I know of a lakeshore lot worth more than three thousand dollars per foot that was created by dredging mucky sand from offshore and dumping it on the adjacent wetland to create a strip of land resembling a natural levee of a swampy floodplain.

One would think that such astonishingly high prices would bring lakeshore development to a halt. On the contrary, it has only raised the demand to increase the supply of waterfront lots. The simplest method is to slice original lots into strips until the cottages are cheek by jowl. Sometimes the lots are only infinitesimally wider than the required frontage, which in many cases is only a hundred feet. In some cases, the original lots were large enough to be carved up into premium lots. Savvy real estate developers have been prospecting for this kind of lakeshore gold for decades. The following real estate advertisement, published in 2007, boasted of a "mother lode" and captures the reality that for some, lakeshore life is mostly about making money.

> Private parties, developers and investors have all recognized that the purchasing of lakeshore property is a good investment that pays dividends well beyond a generous financial return. This recognition has led to a scarcity of premium lakeshore opportunities.[2]

The main sales pitch followed with a description of thirteen prestige lots being carved from an island in Wisconsin's Red Cedar Lake. This used to be one large lot during the Great Depression, on which a wealthy Chicago financier summered with his family. Now it will be used by at least a dozen families, intensifying the impact on the lake and presumably driving the price of nearby real estate even higher. The grass on the croquet lawn advertised as part of the complex, and on the exclusive private golf course to be developed on the adjacent mainland, will probably require pumped water, fertilizer, herbicides, power mowers and leaf blowers, and mosquito zappers. All of these things will either directly leak chemicals into the lake or indirectly generate aerosols that will dust its surface.

Developers have several alternatives to the carving up of large lots. One is to buy an old church camp or summer camp for boys or girls and upgrade it to a full-service, year-round time-share resort for adults. Making new land by filling wetlands is now mostly illegal, done only when legal loopholes can be found, one of which is to "mitigate" the loss by construction of a replacement wetland somewhere else. A recent popular method is to create instant, cul-de-sac neighborhoods of single-family homes and/or condominiums set back from the lake, while retaining the original shoreline lot to provide a common access. In this model, known as keystoning, what may have been a flimsy wooden dock serving a single cottage can become a private marina serving dozens of families, something like the yacht and beach clubs of the Atlantic shore.

For well-heeled investors, a good way to accommodate increased demand for access to lakes is to build in a remote, pristine setting. In this scenario—called leapfrogging—the lower price for more distant land offsets the higher price of constructing everything from scratch, namely utilities, roads, and schools. An additional advantage is that smaller, more rural communities often have less restrictive zoning regulations. With this pattern of development, full-blown lakeside communities, many with upscale houses, can appear almost overnight in places where only a guiding shack or a cluster of cabins stood before. In turn, these developments spawn others farther out, leap by leap.

The concentration of human impact on lakes can intensify even when new domiciles are not allowed. The most common method is to purchase an existing lot with an older, smaller cabin or cottage, tear most of it down, and greatly increase its size, preserving just a tiny fraction of the original house to meet the letter of the law.[3] One creative solution to a shortage of bedrooms is to build an elaborate shed, then finish it off with all the comforts of home, as well as a portable toilet like those used at construction sites and outdoor fairs. Parking a motor home next to the house and plugging it into the house utilities—electricity, cable, Internet—yields extra space as well.

The most intrusive form of development on an existing, built-up lot is to upgrade the dock into another living space, complete with rooms, stairs, patios, chemical toilets, and covered boat lifts. Such personal family wharves block the view of abutting property owners parallel to the shore. When dock upgrades occur on both sides of a property, the lake view from the middle is like that of a horse with blinders.

Ancillary development well back from the shore imperils lakes as well. This usually takes the form of strip malls, schools, and service facilities built along secondary access highways that lie within the lake watershed. Except for those rare cases where the runoff from roofs, parking lots, and roads is artificially drained to a large river or the sea, most of what leaves such interlachen developments as liquid must reach an adjacent lake, either directly as runoff or indirectly as groundwater. This includes runoff not only from precipitation but also from everything rinsed away from the dishwashers and toilets of fast-food restaurants. To whatever extent possible, drainage is directed toward the groundwater system. But even there, polluted water doesn't disappear. It's just diluted, delayed, and improved along the slower pathway. The air pathway is a means of pollution transfer as well.

THE SECOND MEGATREND IS SOCIAL. Kids are less interested in lakes than they used to be, begging the question Who will take care of lakes in the future? Many people who can afford lakefront property are childless. Those who do have children tend to have fewer and tend to be more fearful about their safety, monitoring them more closely in what is now perceived to be a more dangerous outdoor environment, full of creepy crawly things and invisible hazards. The children themselves now tend to eschew outdoor activities in favor of electronic devices ranging from theater-sized, high-definition screens for video games to tiny handheld computers. Do-anything mobile phones, iPods, MP3 players, portable DVDs, social networking sites, and the like mean that a physical presence at the lake need not require a mental presence.

Indeed, fewer people, especially children, have an interest in their natural environment. They are experiencing what author Richard Louv calls "nature deficit disorder." Inevitably, this diminishes their motive to understand, invest in, and protect the natural environment. One clear indicator of this trend is the decade-long decline in the number of fishing licenses being sold in the heartland states, especially to young adults. Many adults now living on lakes have never gutted a fish in their lives. This would have been most unusual a generation ago.

"In the space of a century," Louv wrote, "the American experiences of nature have gone from direct utilitarianism to romantic attachment to electronic detachment . . . Baby boomers—Americans born between 1946 and 1964—may constitute the last generation of Americans to share an intimate, familial attachment to the land and water."[4] My own family reflects that shift. My grandparents grew up as farmers during Louv's utilitarian stage, on my mother's side working with draft horses until the 1930s. The following generation expressed its "romantic attachment" to the land, on my paternal side through the midcentury lake culture and on my maternal side with annual fishing trips to the north. My wife and I were thoroughly imbued with a romantic attachment to lakes in the 1950s and 1960s. Yet while attending our annual lakeside family reunions during the last two decades, even on beautiful days, I have discovered my children and their cousins hiding away in the semidark, mesmerized by whatever electronic gadget is the rage. Once I discovered my son playing a battery-operated fishing game in the shed where we keep our fishing poles.

With fewer kids scurrying about and fewer still popping in and out of neighboring cabins, cottages, and camps, life at the lake has become more adult oriented, and less socially fluid. The seasonal family lake culture I knew as a child remains, but has since become balkanized, based on competing interests.

Users with traditional interests occupy the same cottage, cabin, or camp that their parents did, or have purchased one that is similar. Their focus remains on family and friends—each visit being a mixed bag of food, conversation, lawn games, and low-key water activities like swimming, fishing, and boating. Theirs is a lifelong habit associated with pleasant rather than exciting outdoor memories.

Another interest is to view lake properties as private resorts. Cabins and lakefront facilities are analogous to ski lodges at mountain slopes. Jet skis are optional. A powerful boat for towing skiers in the summer and a snowmobile for racing around on the lake when it's frozen are important possessions.[5] Parties are thrown on pontoons. The old fish-cleaning station has been converted into a three-season sunset gazebo complete with a propane fireplace. The refurbished boathouse has a viewing deck for evening cocktails. Lawns run right to the edge of the water.

There has always been an interest in lakes by the "sportsmen," who emulate the outdoor lifestyles of pioneering trappers and woodland guides. Hunting and fishing are the dominant activities. Fish, fur, feathers, and antlers adorn their cavelike dens. The equipment list includes all-terrain vehicles with gun racks and specialized fishing boats with elevated seats, satellite navigation, and built-in beer coolers. In the opposite direction are those for whom the chief importance in a lake is to fill picture windows with an aesthetic backdrop. This is especially true for retirees from urban areas and working couples too tired to do much except look. Days, even weeks, can go by without them walking down to the shore.

Those with a bent toward natural history and conservation likely keep a pair of binoculars on a peg near the window to identify birds, and probably own cross-country skis. The opposite interest is to view lakeshore property chiefly for its investment value, with buying, selling, and renovating being the principal activities. Many public and private professionals take care of lakes for pay: lawmakers legislate, scientists investigate, regulators monitor, administrators hold meetings, engineers fix, and field officers enforce the law. Native Americans see lakes both in traditional spiritual terms and as a useful resource.

The conflicts are many. Most people have a dominant interest, even if they share all the rest. Those in search of tranquillity and with a sedentary bent are perturbed by the engine noise of motorized waterfront sports. Traditional seasonal family users are put off by suburbanization and the gentrification of lake communities. Conservationists and real estate developers lock horns. Sportsmen and government regulators cross swords on issues of regulation and enforcement. These and other inevitable conflicts reduce the effectiveness of voluntary lake associations because one or more interests can suppress or squelch the legitimate interests of others. When an association is overly committed in one direction, or splintered by competing agendas, it is the lake that loses.

MOST OMINOUS OF ALL MEGATRENDS threatening the future of lakes is the impact of climate change. Residents of the upper Midwest who recall legendary blizzards and temperatures descending to minus sixty degrees Fahrenheit may not see global warming as a bad thing. They forget that the perceived benefits of winter global warming, namely personal comfort and energy savings on space heating, come at the expense of ecosystem health. They also forget that any benefit of winter warming will be offset by the energy-consuming challenges of being broiled during the summer. Record highs for Michigan, Wisconsin, Minnesota, and North Dakota are 112 degrees Fahrenheit or above, all set in 1936, before the well-documented rise in air temperatures beginning in the middle of the twentieth century. When it's that hot, the best place to be is submerged beneath the surface of a cool lake, assuming it hasn't evaporated.

According to the 2007 Fourth Assessment by the Intergovernmental Panel on Climate Change (IPCC) and the U.S. Climate Change Science Program, the glaciated fringe, on average, has warmed approximately one degree Fahrenheit in the last century, with slightly less change to the east.[6] Precipitation has risen as well, with stronger gains in the Northeast and Dakota sectors than near the Great Lakes. Extremes between flood and drought conditions are also becoming more common, there is less snow on the ground, and the season for lake ice is shorter.

These historic trends are consistent with future projections through the end of the twenty-first century.[7] Accordingly, mean annual temperatures within the glaciated fringe are predicted to rise faster than the global average, ranging from five to seven degrees Fahrenheit in the east and seven to nine degrees in the west.[8] The U.S. Climate Change Science Program predicts a strong rise in mean annual temperatures in the west and a slight rise in the east, with winter minimum temperatures rising strongly in both areas. Across the whole of the glaciated fringe, it projects a rise in mean annual precipitation of 5 to 10 percent, with stronger gains in winter than in summer. The largest regional differences involve summer precipitation, which is projected to drop in the west and rise in the east.

The greatest concern regarding individual kettle lakes is not whether the average annual temperature will warm or cool, or whether there will be more or less precipitation annually, but whether there will be more or less water in the lake basins.[9] Extra warmth will likely increase precipitation immediately downwind from maritime sources, while increasing evaporation at greater distances. Hence, kettle lakes and ponds east of the Alleghenies will likely gain water, whereas those farther toward the continental interior will likely lose. Air moving north from the Gulf of Mexico to the upper Midwest will likely drop more of its moisture before reaching the northern heartland.

These predictions by numerical climate simulations match the prehistory told by fossils from the lake archive for the period between about eight thousand to three thousand years ago. During that interval, the prairie-forest boundary shifted eastward and lake levels fell. Most heartland kettles didn't dry out completely, but their shorelines dropped significantly and their flows became more sluggish. Kettles east of the Alleghenies also experienced lower conditions during the middle of that period, though the effects were less dramatic.

Uncertainties associated with regional predictions remain high.[10] But even if they are correct in every case, regional predictions cannot tell us how individual lakes will respond. This is especially true for kettles because their basins are more complex in shape, their groundwater budgets are more sensitive, and their ice cover is more important. For example, a drop in the water surface on a lake with a broad margin of shallow water will reduce the surface area exposed to summer sunlight, perhaps cooling the rest of the lake volume following autumn turnover. Shifts in the timing of ground freezing, snowfall, and spring rain will tip the balance between runoff and groundwater flow, influencing for better or worse the volume, temperature, and chemistry of the lake water. The expected increases in the intensity of storms will likely tilt the water balance toward more runoff, which brings in more pollution, and away from infiltration to the groundwater system, which brings in less.[11]

Changes in forest composition will also lead to indirect changes in the thermal and water budgets of lakes. Most projections are for a partial replacement of conifer-birch forests with more deciduous elements. Deciduous forest captures and holds less precipitation on its branches during winter. This will increase infiltration to the groundwater system, which is good, but will send that infiltration through a soil richer in nutrients, which is bad. The more-intense precipitation events that are predicted will increase runoff, driving up the nutrient flux as well, especially if storms follow periods of drought when the uptake of nitrogen and phosphorus by plants is deactivated.

Of special concern is the impact of the projected water deficit on marshes and bogs in the heartland. They've been called the "kidneys of the landscape" because they sequester pollutants that would otherwise reach open water. For decades they've been sopping up toxic metals, persistent organic pollutants, and phosphate by binding them to organic colloids, and reducing the nitrogen level by converting biologically active nitrogen to a harmless gas. They've also been minimizing flood flows by slowing them over countless tussocks, shrubs, and branches. This slowing effect allows wetlands to trap fine-sedimentary particles onto which nasty chemicals are adsorbed. The concern is that wetland soils will decompose downward, releasing what they've been holding for all these years and raising flood flows.

The hotter the climate becomes, the more attractive local clusters of lakes will become, even if they are fewer in number and lakeshore perimeters are smaller. Right now, places like Chicago experience killer heat waves about once every fifty years.[12] By 2030 this is expected to happen once per decade, and by the 2090s, every other year. Increased summer heat will drive urban residents northward to lakes, intensifying human impact for better or worse. During the winter the ice cover will be thinner and of shorter duration, shifting lake ecologies and making winter activities like ice fishing and snowmobiling less safe and attractive.[13]

CONSIDERING ALL TYPES and sizes of natural lakes in the United States, kettles probably hold the dearest affections of the greatest number of voters; but this hasn't brought them to the forefront of political theater where coastal marshes and scenic rivers enjoy celebrity status. Based on state and federal funding formulas, small lakes in general and kettle lakes in particular seem to be considered the least critical freshwater resources in the United States, virtually insignificant when compared to larger lakes, rivers, wetlands, estuaries, aquifers, and multipurpose reservoirs for flood control, hydropower, and water supply. Unlike individual streams, which link into larger and larger entities and in which downstream residents have a stake in everything that goes on upstream, individual kettle lake communities, especially those without inlet or outlet streams, tend to be small and happily self-absorbed.[14] Each is an island of water and people unto itself. This geography makes kettles ideal places to get away from the hubbub of life but terrible places to develop political solidarity with other lake communities. A voting block of one isn't much of a block.

Indeed, kettles are political Lilliputians when compared to the Great Lakes. Maintaining good diplomatic relations with Canada requires Congress to spend money on these inland seas, especially on Lakes Michigan and Erie, which border the province of Ontario, where the greatest number of Canadians reside. These two land-locked lakes alone claim a de facto voting block of fourteen U.S. senators, dozens of representatives, and scores of federal offices dealing with the sovereign nation to the north. In comparison, there can be hundreds of highly populated kettle lakes within a single U.S. congressional district and thousands in a senatorial district. Of the nineteen states with Laurentide kettles, fourteen share their water resource dollars with either the Great Lakes or the sea. And of the remaining five states—Montana, North Dakota, South Dakota, Iowa, and Vermont—none are popularly known as lake states.

Kettles aren't critical to the national economy or to national defense. They're usually not associated with urban social problems that attract federal revenue streams. With the sole exception of rock star Don Henley's advocacy for Walden Pond, celebrities have not adopted kettles as pet causes, at least not to my knowledge. They lack the drama and toxic-shock imagery that make the nightly television news. Neither do they generate front-page stories by catching fire or causing massive fish kills. They are seldom involved in high-profile environmental issues like scenic river legislation, estuary protection, or wilderness preservation. And now that bald eagles have recovered, their wildlife issues usually involve common local species, rather than charismatic ones like penguins and polar bears. Kettles are sometimes habitats for endangered and threatened species, such as freshwater mussels and other innocuous creatures. To my knowledge no kettle aside from Walden is a poster child for any cause. Most of the damage done to them is done by nice people, making newsworthy bad guys and scapegoats hard to find. And with most of their shorelines in private ownership, there's little that regulators can do except monitor lake status and hope for improvement.

Perhaps because of their political insignificance, thousands of kettles have major problems that will ultimately impact us all. These are being addressed by countless federal, state, and local programs that apply state-of-the-art limnological techniques and involve thousands of trained community volunteers. Not being a professional lake scientist, manager, or administrator, I have little to offer in terms of concrete suggestions to the good work that's already being done. Instead, I offer four broad recommendations for the lake-management community to chew on.

First, caring for kettles would be more effective if we isolated them as a special species of generally small lakes whose archaeology, history, and limnology is distinctive. Small to average-sized kettles lie at one end of a continuous spectrum of lakes, the other end being large bedrock lakes that exchange most of their water via surface inflow and outflow.

Separating kettles from the menagerie would be helped if we quit calling them glacial lakes, as if this were some kind of unique identifier related to their management. No kettle has ever been "gouged out" by the ice, notwithstanding the steady stream of disinformation to that effect. Most are, in fact, sandy sinkholes rather than rock-carved basins.

Kettles also have little in common with the remnants of ice-front lakes or those held up by lodgment till. Both of these types are shallow, underlain by impermeable mud, and have a hydrology dominated by surface streams and evaporation. Moraine-dammed lakes in mountain valleys have more in common with nonglacial artificial reservoirs behind tall dams than with kettle lakes or ponds.

At first glance, kettles look similar to the tens of thousands of small rock-carved and drift-dammed lakes that speckle the inland heart of New England, the Adirondacks, the Catskills, the highlands of northern New Jersey, and the Leatherstocking region of eastern New York State. Thousands more dot the rocky margins of the Great Lakes, especially along Michigan's Upper Peninsula, the northern shore of Wisconsin, and the Boundary Waters Canoe Area in northeastern Minnesota. Such lakes appropriately invite a comparison to kettles based on size and recreational use, but usually not in terms of hydrology and chemistry.[15] For example, the small, rounded glacial lakes called tarns suggest a similarity to kettles, one picked up by Thoreau. But they are the complete antithesis: high in the mountains rather than on lowland terrain, rocky rather than sandy, fed by streams rather than being isolated holes, and tundralike rather than surrounded by farm and forest.

What unites kettle lakes and ponds as a group is not their icy origin but their geographic patterns and material composition. Though the population varies greatly, there is a preponderance of droughty, silica-rich, and generally acidic sand and gravel, through which the sideways groundwater flux is high and the vertical flux through the silt liner is low. When pristine, kettles usually run leaner and cleaner than lakes of similar size with more fine-grained watershed soils with a higher proportion of surface runoff. Unfortunately, their dependence on direct precipitation and groundwater makes them more sensitive to storm-water runoff: The use of kettles as municipal drains is an acute problem in many communities. The more chronic problem involves groundwater pollution, because a higher percentage of their water percolates down through the "coffee filter" of the soil, leaching whatever else is there. They are harder to police because sources of contamination are more difficult to find and regulate. They have longer residence times than normal, which, for better or worse, will delay the onset of degradation but extend the length of time there is a problem.

Kettles also have softer edges. This makes them easier to develop for recreational purposes because excavation costs for roads and building foundations are lower than with either rock or till substrates, and percolation tests for septic system approvals are more predictable and likely to be approved. Because the water levels in classic kettles are controlled by groundwater and because the shoreline materials are more erodible, kettles typically have broader shallows than bedrock or till lakes of the same size. This makes their shorelines and water temperatures more susceptible to changes in water balance; their ecosystems more vulnerable. Slight changes in water level will also exert a proportionately stronger influence on the surface area of adjacent wetlands.

My second recommendation is that we quit forcing this round peg of a lake into the square hole of standard watershed models so often required by federal and state programs. The standard watershed model consists of a branching network of streams within a topographic basin that has a well-defined edge, usually marked by a perimeter ridge. Within each basin is a hierarchical system of tributaries arranged in a fractal pattern not unlike the branches of an elm tree, with Y-shaped forks. This model works well in mountainous regions with steep, rocky slopes and in lowland regions with fairly impermeable soils where the land sheds water systematically. In this linear link model, the most important places are the nodes where two smaller streams merge to form a larger one. Below each node, the properties of the stream water usually reflect the simple sum of the properties from each stream above it.

The math isn't that simple when a node is occupied by a lake, especially if it's a large one, especially if it's a kettle. Rather than acting as a simple junction, the node becomes a physical reservoir that dampens, attenuates, and averages the inflows, and is also a chemical reaction chamber. Molecules falling out of the sky as rain mix with those from multiple incoming streams. In turn, these are mixed with water changed by biological processes within the lake and by groundwater inflows. In other words, the kettles delay, buffer, amplify, and/or cancel the inputs from the upstream links.

Large lakes in flat, sandy terrain reverse the concept of a drainage basin. In fact, they are often runoff-producing islands in a sea of drier ground, capable of generating stronger flows than the land itself. When concentrated within a chaotic sandy landscape—for example, the counties of Vilas in Wisconsin, Itasca in Minnesota, Crawford in Michigan, and Barnstable in Massachusetts—they convert the hydraulic geometry from an easily understood system of linear links to a confusing geometry of polka dots and wiggly lines. Isolated thunder-storms, for which the heartland is famous, can raise the level of a lake surface anywhere within the watershed, often reversing the flow di-rection of streams. Tributary networks aren't supposed to behave this way. More often than not, kettles serve the streams rather than the other way around.

For example, on most drainage basin maps, the headwaters of the Mississippi River are represented as a dendritic system of tributaries that flow through lakes on their way toward New Orleans. From a hydrological perspective it's probably best to think of the area as a collection of independent lakes, the overflows of which just happen to be linked in a downstream direction.[16] Within each lake the river completely loses its identity, mixing and mingling with water from various marshes, bogs, aquifers, and short, sluggish streams. Water entering one side of a lake may spend several years within the basin before joining the outflow to the next lake in sequence downstream.

The traditional linked-line model for watershed management doesn't work in other kettle settings either. On Cape Cod most of the water infiltrates straight down, mounds up above the water table, and migrates sideways to the sea. Large kettle ponds there are simply places where glacial sinkholes intersect the mound. Smaller, usually more elevated kettle ponds there are the sources of streams, rather than the places to which they flow. In prairie kettles, precipitation and snowmelt move mostly straight down, trickle radially toward the base of a till depression, then go straight back up as evaporation. Where's the watershed in that? On elevated kettle moraines like those of eastern Wisconsin and the West Hills of Long Island, the water flows downward to the bottom of the cup created by the silt liner. There the pond surface rises until the excess spills over the edge like a slow-motion underground waterfall.

My THIRD RECOMMENDATION is bureaucratic and managerial. The challenge of preserving and protecting kettle lakes is large enough to warrant the creation of a national center for freshwater research within which small lakes would be central to the agenda. All big federal agencies were born in response to some hot-button issue other than lakes, enabled by an act of Congress, given a mandate, and handed a line-item budget. Their involvement with small lakes is, with one exception, the result of agency evolution, rather than initial mandate. For example, the U.S. Army Corps of Engineers was chartered to build military forts and barricades during the American Revolution, especially as it involved the mouths of navigable rivers. More recently, it has extended its political reach upstream all the way to headwater wetlands, giving short shrift to the many small lakes in between.

The U.S. Geological Survey's Water Resources Division has a well-earned reputation of excellence involving issues of water quality and quantity, but with a mandate focused on streams and rivers, not lakes.[17] The U.S. Fish & Wildlife Service does a fine job managing lakes on its lands as habitat for waterfowl, fish, and terrestrial critters. The U.S. Forest Service and the Bureau of Land Management manage lakes—many of which are artificial reservoirs—as one of many natural resources on blocks of federally owned land. The Centers for Disease Control investigates lakes for their role in contributing to waterborne diseases. With each change in federal administration, the one or more lake scientists within regional and state offices of these and other agencies get shuffled around. Each time, the question What do we do with the lake guy? is asked. Why not take the lake guys from these many agencies and let them work together on larger issues in the same way oceanographers team up to work on their single large places, a model followed by space and climate scientists as well?

Offices within the U.S. Environmental Protection Agency may be the closest things there are to gatherings of federal lake scientists, but they must also work in an agency with a principal focus elsewhere. The EPA was created out of whole cloth in 1970 to police the discharge of pollutants from all sources, especially from politically visible point sources like the smokestacks and effluent pipes of power plants, factories, and municipal wastewater treatment plants. Freshwater is only a small fraction of the EPA's broader mission; lake water is only a tiny fraction of that.[18] Initially the water quality of lakes was investigated, regulated, and funded by the EPA under Section 314 of the Clean Water Act, called the Clean Lakes Program. Funding was discontinued in 1994. Since then regulation has been subsumed and muddled under the more general category of so-called nonpoint pollution, covered in Section 319. As recently amended, less than one percent of the written documentation for the Clean Water Act, which gives the EPA its power, deals with lakes.[19] In terms of budget, it's even less.

The good news is that part of the EPA, albeit a tiny fraction, is working hard to assess small lakes nationally, and that the states are being involved to the maximum extent possible. The bad news is that no more than one out of every five thousand kettles will receive any attention at all, and that there is no guarantee that the program will continue beyond the first sampling, even though repeat samplings following the same protocols provide the best information about lake trends. Also worrisome is that the national lake assessment is embedded within a broader assessment called the National Water Body Survey Program that is mandated to investigate other types of freshwater with more political clout. This water-body survey is a retooled version of its predecessor, the Environmental Monitoring and Assessment Program, which, in turn, had predecessors back to the birth of the agency.[20]

What the EPA has done best is to fund state programs, which have done a pretty good job, depending on the degree to which lakes are important to each state. Vermont has the most focused lake program in New England, concentrating on the state's one large lake, Lake Champlain, although substantial attention is paid to invasive species in other Vermont lakes. The heartland states of Minnesota, Wisconsin, and Michigan have the nation's largest and most comprehensive lake programs, but even there, the Great Lakes get an inordinate chunk of attention.

Realistically, the only way small lakes in general, and kettle lakes in particular, will receive the attention they deserve will be through the strengthening of private, nonprofit lake management associations.

They exist solely to take care of individual lakes, or small lake clusters. There's nothing new about this self-interest model. Some lake associations have been active for more than a century. Through their bylaws and budgets, they can accomplish what agencies cannot, which is to concentrate attention on a single place, rather than many. They can accomplish what individual lake-property owners cannot by themselves, which is to prioritize issues, hire limnologists to help them understand what's going on, then hire aquatic engineers to mitigate problems. They can also retain lawyers and lobbyists to advocate for solutions, employ grant coordinators to obtain funding, retain experts to rebut politically motivated agency-think, and send delegates to regional and national meetings of the North American Lake Management Society. Practically all associations are voluntary, democratic, and governed by an elected board, with members paying dues to protect their own little shared piece of heaven, their single blue dot in an immense galaxy of lakes and ponds.

There are two main problems limiting the effectiveness of lake associations. First, and most important, they usually lack the legal, political, and financial clout to achieve their goals. Legal clout can only come from the top down, meaning federal and statewide zoning regulations that apply what limnologists already know. At a minimum, lakeshore zoning would carefully regulate on-site wastewater disposal, minimize chemically treated lawns, insist on mandatory checks for hitchhiking invasives, require a buffer of wetland plants along the shoreline, prevent the wholesale clear-cutting of weeds from shallow water, require that submerged objects like boulders and logs be left in place for habitat, and limit the impact of powerful boat engines and jet skis. Regulation of pollutants arriving from more distant sources is much more complex, because it involves the whole earth system, or precipitation, surface water, and groundwater fluxes. That requires state and federal involvement.

Some, if not much, of the money needed to monitor, model, and manage lakes must come from user taxes such as fishing and boat license fees and owner taxes based on the value of shoreline property.

More needs to come from the state and federal governments, not necessarily as new tax revenue, but gleaned from built-in agency inefficiencies and lumped into a pool of funding accessible to individual lake associations. With such a pot of money, they could contract for limnology services that would include installing a monitoring system that involves community participation, running simulation models for different scenarios that make optimum use of the present scientific knowledge, experimenting with different control methods, and making recommendations for engineered solutions. This "think locally, act locally" model works especially well for isolated lakes because property values are strongly correlated with their ecological health.

A second problem involves power sharing between stakeholders. Members of lake associations are usually property owners, those with the largest stake in the ecological health of a lake. Often they are legal residents of other towns, cities, counties, states, and sometimes nations. At the same time, the lake they care about above all others is subject to regulations implemented by someone else, in this case by

*Limited shoreline development in the Litchfield Hills, Connecticut.*

local and state residents, who may have only a distant and abstract view of the situation, even though in some rural towns, lakeshore revenue provides the largest single chunk of their operating budget. Out-of-state or out-of-town lakeshore property owners are thus taxed, but are not eligible to vote, a situation bordering on taxation without representation. Conversely, in the case of state funding for a lake that is largely privately owned, those who pay state income tax may begrudge funding for what amounts to a private community. These built-in conflicts create tension that makes the need for lake management associations all the more important, because they focus on the natural entity of the lake system, rather than the artificial entities associated with governments and bureaucracy. Every stakeholder, ranging from the occasional out-of-state visitor to the mayor of a lakeshore city, has a vested interest in the health of a community's lakes.

GIVEN THE PRESENT MOOD of the taxpayers for smaller government, the graying of the lakeshore population, the balkanization of lake culture, and the nature deficit disorder among young people, perhaps the best thing that could be done to improve and protect our nation's small lakes would be to take a more creative and intensive approach to public education. Lake scientists already participate in school-based and summer-camp programs. A wonderful example is the Maine Lakes Conservancy Institute, a nonprofit educational provider that runs outdoor classes for middle school students on pontoon boats. These are lucky students. Most, especially in inner cities, never get such an opportunity. So much more needs to be done to incorporate lake science into elementary and secondary school curricula. Oceanography is glamorized. Why not limnology?

Even when lake science is offered as part of school curricula, it often misses the boat with its overemphasis on "pond life," meaning tadpoles, lily pads, and the like. Limnology was indeed founded by zoologists, and that tradition will hopefully continue. But what about the kids who might be more turned on by the physics, chemistry, and climatology of lakes? How about those who might find the fifteen-thousand-year-long history of American kettles almost as interesting as dinosaurs and volcanoes? How about those high school students for whom the sociology of lake life is most intriguing? As an integrative school topic, small lakes are not only the right scale to interest a classroom full of kids, but they also have the advantage of spanning all the curriculum subjects. They offer what may be the best chance for students of the heartland to forget about the disciplines being imposed on them by adults and pursue their interests broadly.

Adult education about lakes tends to be problem specific, often too narrowly targeted toward the main issue on the table, rather than the broader systems approach that would help them make more effective decisions. Limnologists routinely work with adults, but usually as technical consultants or managers working for lake associations. Perhaps greater attention should be given to adult educational programs about how lakes actually work, rather than what to do when a problem emerges, if only because it will help them appreciate what they already love even more.

ANTHROPOLOGIST RICHARD LEAKEY reminded us with the title of his book that we are all "people of the lake" in one way or another.[21] Without freshwater pools and ponds, our vertebrate ancestors would not have survived their passage up from the sea to claim higher ground. Throughout human evolution, watering holes on the savannah and lakeshores in rift valleys have been sites of both opportunity and danger. For the indigenous peoples of the glaciated fringe and for the colonial and American settlers who came later, they were the sites of special resources. After the Civil War, and especially after the 1930s, they became sources of ecological, recreational, and spiritual delight. In the not-too-distant future, concerns about freshwater will rise to the level of concerns for energy today. In such a world, where every drop will count, the band of kettle lakes and ponds crossing the northern United States will be seen as a shared national resource deserving of more careful monitoring and vigilant protection.

Getting to know the Great Lakes is like getting to know a family of giants. Getting to know kettles is like getting to know a whole nation of people. This can be achieved only by exploring their common history and behavior. All were born during the meltdown of stagnant glacial ice more than ten thousand years ago. After a few millennia of being dominated by physical forces, these aquatic systems stabilized to become finely tuned ecological machines buffered against changes in climate, forest processes, and indigenous human use at the scale of decades and centuries. The Ojibwe, and presumably the Archaic Algonquins who preceded them, believed that each member of their tribal nations was like a single wave upon a universal lake, a transient pulse of local energy on something larger and more mysterious than themselves.[22]

Euro-Americans would have done well to follow the native example. Instead, their social system encouraged them to exploit lakes for economic purposes through the fur trade, farming, mining, and logging. This was followed by a century when lakes were being transformed into recreational resources. Many lakes are still fine. Others were pushed beyond the limit of tolerance. With the demographic, climatic, economic, and political changes looming on the horizon, the time has come to rededicate ourselves to the lakes that played such an important role in our history, and which have helped so many families learn to appreciate nature.

When my first grandchild catches a sunfish off the end of the dock at Union Lake, Minnesota, the level of lead neurotoxin in its tissue will reflect five generations of unintentional impacts by our family and those of neighboring cottages. There will be the lead from the gasoline I burned in my Chevy Biscayne to get there during the 1960s, the shotgun pellets my father blasted out over open water when he went duck hunting in the 1940s, and the lead solder used to plumb the kitchen after my grandfather bought his cottage in the 1920s. This three-generation dose does not include the lead leached from paint on the old boathouse during my children's generation in the 1980s, nor the aerosols that will drift in from Dakota farm fields being plowed under for ethanol production during the present millennium.

Beneath this legacy of invisible harm is an even greater legacy of family joy; not only for our family, but also for the millions of other families whose favorite lake has filled their lives to the brim. My hope is that this joy will empower individuals, teachers, shoreline neighborhoods, lake associations, and governments of all sizes to protect and restore our nation's small lakes beyond Walden, from Maine to Montana. No kettle need become a "tragedy of the commons" because someone wasn't paying attention or because the neighbors couldn't agree on what to do.

# Acknowledgments

My first book on signature landforms explored New England's fieldstone walls. Though my parents were proud of this effort, they didn't quite understand why anyone would write a book about old fence lines. So I decided to write about a signature landform that they—and millions of others—care about greatly, the galaxy of small, sandy lakes so important to residents of the northern United States. Thanks, Mom and Dad, for imbuing me with family lake culture as a child, and for moving us north to live in lake country year-round.

As this book took shape, I began teaching an interdisciplinary honors course titled *Walden* and the American Landscape. Over time, I became especially grateful for the thoughts and words of Henry David Thoreau and to those he inspired. My co-instructors—professors Wayne Franklin in English and American studies, Robert Gross in New England history, and Janet Pritchard in art and art history—and our eager students helped frame my thinking about why kettle lakes and ponds were important to American history and culture.

Much of the early research for this book was done while living in a lake cabin on the shore of a Minnesota lake near the Mississippi headwaters. The manuscript was written on a salty island cove in Narragansett Bay, Rhode Island, where hundreds of kettles lay within easy reach, especially those of the Charlestown Moraine, Cape Cod, Block Island, and Martha's Vineyard. I thank the Department of Ecology & Evolutionary Biology and the College of Liberal Arts and Sciences at the University of Connecticut for funding my sabbatical, and my wife, Kristine, for understanding the need for a writer's retreat.

Most of my kettle experience within the glaciated fringe involved Minnesota and New England, two regions of about the same size.

I thank those in between for the hospitality during our many car trips back and forth. To help fill in the blanks between East and West, I consulted the staff at nineteen state geological surveys. Thanks to Laurence Becker, Bob Bergatino, Jon Boothroyd, Gordon Connally, Brandon Curry, Gary Fleeger, John Hill, Howard Hobbs, William Kelley, Linda Laplante, Stephen Mabee, Lorraine Manz, Ed Murphy, Rick Pavey, Mike Prentice, Jean Prior, Deborah Quade, George Springston, Scott Stanford, Dennis W. Tomhave, John Wattig, Thomas Weddle, Lee Wilder, Steve Wilson, and Ron W. Witte. I also thank limnologists Donald Cloutman (Minnesota), James Kitchell (Wisconsin), Chuck Lee (Connecticut), and Amy Samagla (New Hampshire) for their insights. Specialists Byron Stone (U.S. Geological Survey), Anton Truer (Ojibwe language), and Carolyn Freeman Travers (Plimoth Plantation) helped with specifics.

Discussions with ecologist Eugene Likens and limnologist Ken Wagner helped frame the direction of this book. Ken also reviewed the final chapter on lake futures. Colleagues at the University of Connecticut, notably social historian Robert Gross, fisheries biologist Eric Schultz, anthropologist Dennison Nash, and limnologist Peter Rich, offered ideas for sections of the text. Off-campus readers were Sherman Clebnik and George Knoecklein, who read and reviewed a first draft. Thanks also to Chelsae Becce, Michael Forbes, Ellen Kisslinger, and Rochelle Skinner for odds and ends.

This book reaches back forty years to my undergraduate days. Thanks to my college buddy and traveling companion, Harlan J. Brooker Jr., who accompanied me on innumerable lake sojourns. At Bemidji State College, Professor Robert Baker introduced me to aquatic ecology and conservation; professors Jim Elwell and Robert Melchoir to geology; and Harold Borchers, Adelle Elwell, James Ludwig, and Walter Wanek to zoology, biochemistry, general biology, and plant taxonomy, respectively.

Finally, I thank my agent, Lisa Adams of the Garamond Agency; my publisher, George Gibson at Walker & Company; and especially my editor, Jacqueline Johnson, for her translations and transformations.

# Glossary

This glossary is restricted to words that are needed to understand the kettle lake phenomenon, especially as treated in this book. Definitions are informal and followed by a comment about why each term is important.

AQUIFER. An underground body of fractured rock or unconsolidated earth materials that readily holds and transmits water. Most kettles lie within near-surface aquifers composed of sand and gravel.

ARCHAIC PERIOD. An interval of time used by North American archaeologists dating between about 10,000 and 2,500 years ago. It spans the vast majority of time between formation of kettle lakes and the arrival of Europeans.

BASEMENT. The continuous mass of ancient crystalline rocks forming the earth's continental crust. It is widely exposed in the source area of the Laurentide Ice Sheet and was therefore responsible for producing much of the sand.

BOREAL. Cold, generally continental conditions of the northern United States and southern Canada. The term is most widely used to describe a mixture of spruce, fir, pine, birch, and beech trees and cold-tolerant herbs and grasses.

CANADIAN SHIELD. A geographic area where ancient crystalline rocks are exposed over most of central Canada, extending into the United States in northern Minnesota, Wisconsin, and Michigan. The rocks date mostly from the first half of the earth's history.

COULEE. A long, trough-shaped valley usually formed by the overflow of a glacial lake. These were the main canoe routes for the seventeenth-century explorers of the Louisiana Territory.

CRATON. The stable core of a continent consisting of ancient, highly deformed crystalline rock overlaid in many places by less-deformed sedimentary rocks. It excludes mountain systems and coastal plains.

**DETRITUS.** A layer of organic matter on the bottom of a lake or pond, dominated by fragmented plant matter and the remains of plankton. Synonymous with muck, but with a connotation of being more fibrous, and located in shallow water.

**ERRATIC.** A conspicuous, usually solitary glacial boulder brought in from afar and let down on the land gently during ice melt, usually identified by large size or exotic composition. They are especially common on the shores of kettles.

**FAULT.** A planar fracture within or between rock types on which movement has occurred. Because glacier ice can be considered rock, its base can be considered as a special, well-lubricated type of geological fault. Near the edge of the glacier, smaller thrust faults caused by compression were responsible for building moraines.

**FORAGING.** One of three main means of acquiring food within a human ecosystem, characterized by seasonal movements over a landscape. Synonymous with hunting-gathering, foraging was the dominant means of subsistence in the lake-forest-wetland system.

**GLACIATED FRINGE.** Synonymous with fringe. A geographic region lying north of the limit of the Laurentide Ice Sheet and south of the Canadian Shield, where rock outcrops are common. The main concentration of kettle lakes occurs where the dominant surface materials are sand and gravel deposited by meltwater streams.

**GROUNDWATER.** Water beneath the land surface trapped within pore spaces, voids, and fractures, and flowing in aquifers. Of the three main water sources of lakes (groundwater, streams, and direct precipitation), groundwater is disproportionately important in kettle hydrology.

**HEARTLAND.** Synonymous with northern heartland and kettle lake heartland. Refers to the central and northern parts of Minnesota, Wisconsin, and Michigan, the three states where the phenomenon of kettle lakes reaches its culmination. Southern Ontario would be included were the term not restricted to the United States.

**ICE DOME.** A broad mound within an ice sheet and the local center of outflow. The Laurentide Ice Sheet consisted of coalesced domes, two of which were important in producing the band of kettle lakes in the United States.

**KETTLE.** A topographic depression produced by the melting of a stagnant block of ice and the subsequent downward and inward collapse of material. When dry, they are kettle holes, hollows, or valleys. When occupied by wetlands, they are bogs, swamps, and marshes. When they hold standing water, they are pools, ponds, and lakes.

**KETTLE MORAINE.** Any moraine in which kettles are common. Most form when lobes of ice advance and stagnate, leaving slabs and blocks of ice that are buried by copious sand and gravel. Most are low, but conspicuously elevated, ridges of chaotic topography. Synonymous with elevated bands of ice stagnation terrain.

**LAKE-FOREST.** Composite informal name used for the ecosystem of the kettle lake heartland. Short for lake-bog-marsh-swamp-forest.

**LIMNOLOGY.** The science of lakes. Similar to oceanography, but focused at a smaller scale and with different species and techniques. A combination of geology, physics, chemistry, and biology.

**LOBE.** A distinct mass of glacial ice within a topographic depression and/or between highlands at the scale of miles or larger across. Moraine formation, meltwater sedimentation, and kettle formation are concentrated where lobes separate or collide. Lobes occur within sectors and often have sublobes.

**MORAINE.** Any ridge built at the edge of the ice sheet, generally consisting of material thrust up by movement or deposited near the edge, and usually marking a place where the outer margin stabilized. The terminal moraine marks the outermost advance. Recessional moraines mark places where the ice sheet remained in one place or readvanced.

**MUCK.** The layer of organic matter lying at the bottom of a lake or pond, dominated by fragmented plant matter and the remains of plankton. Synonymous with detritus, but with a connotation of being softer and in deeper water. Muck provides the archive for pollen studies.

**NUTRIENT.** An element or substance vital for biological growth. Freshwater macronutrients are phosphorus, nitrogen, and carbon in decreasing order. Many others are vital and are considered micronutrients.

**PALEO-INDIAN PERIOD.** An archaeological interval of time dating from about 13,000 to about 10,000 years ago, which coincides with the ice-age extinctions and major ecological adjustments.

**PARTIAL KETTLE.** A kettle in which the depression was formed by collapse along only part of the margin. Basins in till or bedrock merely dammed up by meltwater sediment do not qualify. Many moraine-dammed heartland lakes are partial kettles. Partial kettles are common where ice blocks in bedrock tributary valleys are buried by sedimentation in the main valley.

**PLATFORM.** That part of the craton (the stable core of a continent) overlaid by slightly deformed sedimentary rocks (limestone, shale, sandstone), rather than by the harder rocks of the shield. Glacial erosion of the platform usually resulted in more loamy, clay-rich tills and a lower concentration of meltwater sand.

**SECTOR.** One of five large geographic areas along the southern margin of the Laurentide Ice Sheet: the Northeastern, Great Lakes, Superior, Dakota, and Montana sectors. All were characterized by a different style of kettle lake formation.

**SILT LINER.** The deepest and oldest layer of lake sediment within kettles, deposited after the collapse of more stony and pebbly material, but before the deposition of organic muck (detritus). Though silt is the dominant sediment texture, clay and sand are also important and sometimes dominant. The lasting hydrologic significance and the impermeability are suggested by the word "liner."

**WOODLAND PERIOD.** An interval of time used by North American archaeologists beginning sometime between 2,500 and 3,000 years ago and extending to European contact. It is associated with pottery and cultivation.

# Appendix A: Kettles State by State

Nineteen states from Maine to Montana are blessed with kettle lakes and ponds created by the Laurentide Ice Sheet. This appendix provides a brief sketch of the geological setting of kettles within each state of the glaciated fringe, arranged in alphabetical order. They illustrate the variety of geological, ecological, historical, cultural, and recreational aspects of the kettle lake phenomenon.

## CONNECTICUT

Except for in its largest valleys, the surface soils of the Constitution State are dominated by sandy till. Hence, there are no large concentrations of kettles. However, the general roughness of the topography caused by thousands of valleys cut to a depth of several hundred feet caused widespread stagnation of the ice, producing one or more undistinguished kettles in practically every town. One of these is **Linsley Pond**, considered by many to be the birthplace of limnology and ice-age climatology in North America. Elongate valley kettles and kame terraces are common in the Quinebaug, Shetucket, Farmington, Connecticut, Shebaug, and Quinnipiac river valleys. The strangest Connecticut kettle of my experience is **Great Pond** in Glastonbury. Due to its unusual plumbing, the water level rises and falls as much as thirty feet on a seasonal rhythm, and with a more erratic rhythm caused by summer rainstorms. Great Pond is a deep dimple in the sandy kettle moraine that dammed up the Connecticut River to form Glacial Lake Hitchcock, on which the most important cities of western New England sprawl from Middletown, Connecticut, to Saint Johnsbury, Vermont. The broad terraces on both sides of this ribbon-shaped lake form a line of kettles nearly all the way to Quebec.

## ILLINOIS

The distribution of kettle lakes within the Land of Lincoln has the shape of a giant bite taken out of the northeastern corner of the state. The southern third of the state and its western half were not covered by ice during the last glaciation. Most of its northern border with Wisconsin is driftless as well. Between Lake Michigan and the glacial limit, however, is a group of a dozen prominent end moraines separated by the clay-rich soils of expanded ice-front lakes. Kettles are most highly concentrated in northern Cook County west and northwest of Chicago, where the south-flowing Lake Michigan lobe turned a corner to the west, concentrating and elevating the moraines. **Bullfrog Lake**—now part of the Cook County Forest Preserve District—is a fine example. Nearby **Volo Bog** is the state's only quaking bog with an open-water center. This was the site of late-1960s environmental activism, where a "Save Volo Bog" campaign was successful. Today it's a National Natural Landmark with more than eleven hundred acres.

## INDIANA

The Hoosier State may be the most underrated state with respect to the abundance and beauty of its kettles, perhaps because it lies in the tourist shadow of Michigan. Small kettles are sprinkled here and there on dozens of moraines that cross the northern half of the state. They are especially common in two interlobate settings, one between the Lake Michigan Lobe and the Huron-Saginaw Lobe, and the other between the Huron-Saginaw Lobe and the Ontario-Erie Lobe. The most spectacular kettles are in **Chain O' Lakes State Park,** located near Fort Wayne. There eight beautiful small lakes lie within a subglacial valley carved by subglacial floods and later occupied by a system of eskers. Further retreat left large blocks of stagnant ice between the eskers, which later melted. A much less spectacular example lies in Elkhart, Indiana. Less than a century ago, **Boot Lake** was a popular site for swimming, boating, and fishing. Between 1972 and 1994, however, it was politely called a "sludge farm," where the city dumped solids after sewage treatment. In 1994 it was reconditioned by dredging and the removal of more than seventy thousand pounds of illegally dumped trash. Later it was converted to an environmental education center. It's now a cause célèbre for kettle lake remediation.

# Iowa

The Hawkeye State wins the prize for the greatest kettle lake hyperbole. In western Iowa, just south of the Minnesota border, is a chain of beautiful, but perfectly average, prairie kettles known as "Iowa's Great Lakes." Residents are well aware that the nearest truly great lake is Lake Michigan, more than a hundred miles away. The largest in Iowa is **Big Spirit Lake**, which is less than one tenth of 1 percent the size of the smallest Great Lake, Lake Ontario. Nearby **West Okoboji** is the deepest lake in the state at 140 feet. Longest is **East Okoboji**, which has the shape of a slackwater river and a depth of no more than 10 feet. All lie within a kettle moraine called the Bemis Moraine, the outermost of the western Des Moines Lobe, which squirted down from through the Red River valley between Minnesota and North Dakota. Lakes are present on other Iowa moraines as well. The Nature Conservancy selected **Arend's Kettlehole** in the Freda Haffner State Preserve as one of the state's finest classic kettles, with its steep sides and nearly twenty-five feet of sediment recording the last twelve thousand years of history, marked by a dry interval midway through the present interglacial period.

# Maine

Kettles in the Pine Tree State have the distinction of being the most overwhelmed by grander landscape elements. The interior is a spectacular quasi wilderness of cold forests, low mountains, and ragged, rocky lakes. The southern coastal part of the state lay beneath a fast-moving tongue of a marine ice sheet, the marginal ice of which floated away as bergs, rather than stagnating in place. Exposed there is either scoured rock similar to that of the Canadian Shield or a thick veneer of mud deposited in slate gray, brackish waters. Maine's largely unknown kettles are common in several blotches of ice stagnation deposits left where the glacier stabilized. **Schoodic Lake** is a beautiful, crystal blue kettle located in a moraine where the marine edge of the ice sheet became grounded on land. The surrounding chaotic, boulder-studded terrain presented great challenges to early mapmakers of the United States, more so than in the mountains to the west. Its most famous visitor was Jefferson Davis, soon to be president of the Confederacy. He traveled there in 1858 to visit an old friend responsible for surveying the baseline for the United States Coast Survey, and to recover from a variety of ailments in the fresh mountain air. Davis left his slaves at home. Maine's most important concentration of kettles, however, lies in the southwestern corner of the state near New Hampshire where ice retreated on land, rather than in the sea. The flood of meltwater sand and gravel draining from New Hampshire's highest peaks flowed down the Saco River valley, burying countless ice blocks, one of which became **Bonny Eagle Pond**. The grandest lake in this region, Sebago, is not a kettle. But this rock-carved basin was dammed to the south by a kettle moraine containing the **Otter Ponds**, a cluster of beautiful, circular kettle lakes used by a YMCA–United Way day camp for the kids from the Greater Portland area. The cool, clear groundwater they swim in comes from the much more magnificent, rock-carved Sebago Lake to the northwest.

## MASSACHUSETTS

The Bay State has **Walden Pond**, the most famous kettle in the world. It was the crown jewel of the Concord lake district that inspired the transcendentalists of the mid-nineteenth century. Surrounded by pine and oak woods as part of a state preservation, it stands in remarkable contrast to nearby **Jamaica Pond**, arguably the nation's most urbanized kettle. It lies in the heart of Boston on Jamaica Plain, the outwash fan that buried a large block of ice here about fifteen thousand years ago. Rimmed by concrete pedestrian pathways, this has been a popular place for swimming, skating, and sailing since the Gilded Age, when famous architect Frederick Law Olmsted was a familiar visitor. Though only sixty acres in surface area, the pond is remarkably round and surprisingly deep, reaching ninety feet, which accounts for its clarity, even in the midst of a great city. Kettles to the north are sprinkled throughout the well-preserved historic landscapes of Groton, Andover, Dunstable, and other towns near the New Hampshire border where ice blocks were flooded with crushed-granite sand from New Hampshire. To the south lies Cape Cod and the Islands, where practically every body of freshwater is a kettle or partial kettle. To the west, kettles flank both sides of the Connecticut River valley in the so-called Five Colleges area. **Factory Hollow** in

North Amherst, like many New England kettles, was raised by a dam to regulate the flow of water to a Yankee mill.

## MICHIGAN

Michigan consists of three peninsulas formerly surrounded on all sides by large glacial lobes. Moraines, sand plains, and belts of ice stagnation topography occur above the lowlands, where flat plains from muddy ice-front lakes are present and, below the highlands, where bedrock lies near the surface. Thousands of kettles, some of the finest in the world, are distributed across all three peninsulas. The Upper Peninsula, jutting east of Wisconsin, is mostly rock, till, and bog, though many fine kettles are present. **Sunday Lake**, near Wakefield, is particularly accessible. The Lower Peninsula is known as Michigan's mitten, with the "hand" peninsula lying west of Green Bay and its "thumb" peninsula lying to the east. Within the hand is Kettle Lake Elementary School in the town of Alto, perhaps the only school in the nation named for this landform. Go Cougars! The waters of nearby **Thumb Lake** are renowned for their turquoise color, their depth of more than 150 feet, and a cold-water game fish known as splake, a hybrid cross between lake trout and brook trout. In Michigan's "thumb" is heavily developed **Elk Lake**, its shoreline almost completely surrounded by year-round and summer homes and lawns that reach right down to the water. It's popular because it lies within easy reach of Detroit, Flint, and Port Huron.

## MINNESOTA

The North Star State is known as the "land of ten thousand lakes," most of which are kettles. The central part of the state has been beaten into submission for nearly 3.5 billion years, forming the nation's only three-way continental divide: north to Hudson Bay via Lake of the Woods, east to the Atlantic via Lake Superior, and south to the Gulf of Mexico via the Mississippi. No other state can make such a claim. Its clean sand came from the granitic rocks of the Canadian Shield. Its stagnant ice came from the multiple lobes and lobe collisions on relatively flat land. **Lake Itasca** is a site of international significance as the source of the greatest river in the United States, the Mississippi. **Leech Lake** is the largest and most chaotic, an assemblage of tunnel valleys, kettles, and moraines right in the heart of the northland. **Union Lake**, in Erskine, is my favorite, simply because it was the first one I got to know. It lies in a kettle moraine only a few miles from the greatest lake of all time, Glacial Lake Agassiz.

## MONTANA

The Treasure State has two distinctions of note with respect to kettles. First, it contains the westernmost kettles of the Laurentide fringe. Parts of the **Lonesome Lake** wetland complex near Havre qualify as partial kettles; a large concentration of teepee rings occurs nearby. Montana's second distinction is to have kettles created by a completely separate glacier system, the Cordilleran Ice Sheet, which oozed southeastward from the Canadian Rockies of Alberta to meet the edge of the Laurentide farther north. (A few kettles from another glacier system, the mountain glaciers of the Adirondacks and northern Appalachians, also formed in the United States, but they were completely overwhelmed and destroyed by the Laurentide advance.) Toward the eastern part of the state lies **Echo Lake**, an appropriately circular kettle within a cluster near the state line with North Dakota.

## NEW HAMPSHIRE

Lumpy granite boulders litter the landscape, and crushed-granite sand fills the bottoms of nearly every broad valley. This state is distinguished by having the headwaters of New England's most extensive meltwater pathway, a horn-shaped system of subglacial channels that opens southward toward Cape Cod and the Islands. From the White Mountains, sand flowed southward in subglacial torrents following the trend of the Merrimack River through Concord, Manchester, and Nashua. From there it crosses the state line toward Walden Pond in Concord, Massachusetts, continues south to Plymouth County, then finally ends at Nantucket Island. Kettles are present throughout the region in places where sand could accumulate. **Chocorua Lake** is the most famous of many kettle lake reflecting ponds near Conway that magnify the White Mountains, the highest cluster of peaks in the northeastern United States.

Nearby **Pea Porridge Pond** is another. These lakes are unusually pure due to the filtering effect of the clean granite sand, and their small beaches unusually sparkling owing to the abundance of bronze mica. **Mirror Lake**, into which drains the Hubbard Brook Experimental Forest, is one of the best-studied kettles in the world. Another interesting kettle has filled with vegetation to become **Spruce Hole Bog**, located several miles southwest of Durham. It's so well preserved and rich in unusual plants that it was named a National Natural Landmark in 1972. Muddy-shored and deep-bottomed **Eagle Pond** near New London helped inspire Donald Hall, poet laureate of the United States.

## New Jersey

Only the northern part of the Garden State was glaciated. Within the New Jersey Skyland terrain of the high Alleghenies are dozens of lakes carved out of Appalachian bedrock, giving them genetic affinity to the lakes in Maine, the Adirondacks, northern Michigan, and northeastern Minnesota. In the highlands east and west of Passaic, however, are moraines curving like a "J" toward Delaware Bay and like a backward "J" toward the Hudson River, respectively. The hundreds of kettles on these moraines are mostly too small to be named. But within Sussex County are kettles large enough to have names. **White Lake, Lake Grinnell,** and **Turtle Pond** are a few good examples.

## New York

The Empire State is the sum of several distinctive terrains, each with a separate version of kettle lake formation. Long Island and Fisher's Island are sandy kettle moraines flanked by outwash plains not unlike those of Cape Cod. **Lake Ronkonkoma** is their grandest kettle, located on the boundary between Suffolk and Nassau counties. Now fully suburbanized, this circular, spring-fed lake was formerly a rural resort town with dozens of hotels where residents puzzled over the strange changes in water level. Before that, it was a well-known point of rendezvous for coastal Algonquin tribes. Densely urbanized New York City has **Eibs Pond**, located in the borough of Staten Island. Though only three acres, it is the largest kettle pond in the city and is accessible by public transportation. **Lamoka Lake** is located in the Valley Heads Moraine, which is responsible for damming up the Finger Lakes across the "hand" of central New York. It may be the most important kettle for archaeologists, being the reference site for the Archaic period in the United States. Kettles also dot the lowland terrain surrounding the flanks of the Adirondacks, where bedrock lakes predominate, as well as along the southeast shores of Lake Erie.

## North Dakota

Differentiating kettles from other types of small lakes is most challenging in the Peace Garden State. This is the heart of the prairie pothole region, where isolated marshes, ponds, and small lakes are gems on the otherwise vast rolling prairie, making them critical habitats for waterfowl, especially during migrations. The deepest and most persistent prairie potholes are usually kettles. Deepest of these is **Kettle Lake**, not to be confused with the one by the same name in South Dakota. It is an exceptionally important archive for the natural history of the Great Plains, containing nearly sixty feet of lake sediment that record postglacial climatic and biological events. During a prolonged drought about six thousand years ago, the waterfowl became so concentrated that their fecal material was transformed into layers of mineral phosphate. On the Missouri Couteau lies the densest cluster of kettles. There is a cluster in Dawson, just north of Interstate 94 where **Mud Lake** is curiously larger than **Big Muddy Lake**.

## Ohio

The heavily industrialized and intensively farmed Buckeye State has three bands of terrain parallel to the southern shore of Lake Erie. The southeast third of the state is thoroughly Appalachian in flavor, with no glacial influence whatsoever. There hills and hollers resemble those in West Virginia. The northern band is a flat glacial lake plain, especially broad toward the west. The central band is a belt of curved moraines formed by lobes that pushed southward from Lake Erie, especially in the Miami Valley. Upon and between these moraines are kettle lakes, most of which have been heavily impacted by human activities. A large, ragged kettle named **Indian Lake**, once visited by Daniel Boone, is its most popular. It was raised with a dam in the 1850s to supply water for a canal system that was later abandoned. The reservoir, with twenty-nine miles of shoreline, became a recreational area known as the "Midwest's million dollar playground." Ohio has only one naturally vegetated and undisturbed kettle pond: **Calamus Swamp** near Columbus. This kettle, managed by the Columbus Audubon Society, is dominated by green ash, American elm, and maples, a southern counterpart to the piney kettles of the North.

## Pennsylvania

Pennsylvania has kettles only in its corners. To the northwest, it shares an elongate cluster of kettle lakes on the south shore of Lake Erie with New York State to the east and Ohio to the west. There the crown jewel of Pennsylvania's glacial lakes is an isolated kettle named **Lake Pleasant**, located near Arbuckle. It is noteworthy because its water is highly alkaline, due to groundwater seepage through limestone sand, giving rise to a lake of exceptional clarity for such a highly developed region, and wetlands dominated by unusual fens. Twenty-four species of plants are listed as species of special concern, and three of its fish species are rare. **Lake LeBoeuf** was visited by George Washington in 1753 when he was a young soldier during the run-up to the French and Indian War. At the time, he was a major in British uniform carrying a letter from the governor of Virginia warning the French to abandon their fort on this strategic overland route between the Ohio River and Lake Erie. This spring-fed kettle was named by the French for the native woodland bison inhabiting the area. Pennsylvania also has kettles to the northeast, flanking the Hudson Lobe. A small cluster in Priceville, near the Delaware River, is an easy day trip from urban and suburban New York, New Jersey, and Philadelphia. Lakes there, especially **Price Pond**, are host to exclusive summer camps for children from families of the eastern megalopolis.

## Rhode Island

The southwestern edge of the Ocean State is a land of straight, sandy barrier beaches and long tidal ponds. These are backed up to the north by a bumpy kettle moraine contemporaneous with those forming Cape Cod, but which is much more bouldery. This Charlestown Moraine is home to the Kettle Pond Visitors Center of the U.S. Fish & Wildlife Service, one of its regional headquarters. Here, and probably only here in the state, kettles command more attention than the nearby sea. Its small museum exhibits and illustrates the local version of kettle formation. **Ell Pond,** a nearby kettle managed by the Nature Conservancy, is a favorite among naturalists for its botany: Bone-dry granite ledges and enormous boulders are juxtaposed against cypress and red maple swamps. Rhode Island also has Block Island, an offshore, upside-down, V-shaped interlobate moraine marking the outer ice limit. There the island charm—irregular hills, bouldery pastures, stone walls, and kettle ponds—results from ice stagnation topography. **Rodman's Hollow,** a local nature preserve, has the dubious distinction of supporting the only naturally breeding population of the American burying beetle (a federally protected species) east of the Mississippi River. Located on the Atlantic flyway within a mile of the open Atlantic, this kettle is a favorite for bird-watchers. Aptly named **Fresh Pond** is its most famous water-filled kettle.

## South Dakota

Kettles in the Mount Rushmore State are fairly small and widely scattered throughout the eastern third of the state. The pattern there is similar to that of Illinois because the western part of the state remained ice free as the eastern side was smeared by clay-rich tills. South Dakota is distinguished by having **Kettle Lake,** the most heavily militarized kettle, at Fort Sisseton, a National Historic Landmark now restored as a museum. Built in the wake of the Sioux Uprising of August 1862, the fort served as a military outpost during the last few decades of the Indian Wars. The high ground of the moraine provided a good view, the local kettle provided a secure supply of water, there was clay for bricks and lime for mortar, and a local stand of trees provided fuel and timber. Another kettle is **Medicine Lake** near Codington, which may be the most alkaline kettle in the state.

## Vermont

The joke about the Green Mountain State is that all the land is a sloping hillside. Almost, but not quite. Running the full length of its western and eastern borders are flat terraces built into ribbon-shaped glacial lakes, Glacial Lake Hitchcock in the Connecticut River valley and Lake Champlain on the New York border. Within the interior there are only a few kettles, even fewer of them of the classic variety, owing to the absence of lowland basins large enough to capture enough sand and gravel to make them. **Clear, Perch, Zack Woods,** and **Collins Pond** in North Wolcott are particularly fine examples of Vermont Kettles. **Kettle Pond** in Groton State Forest lies on the southwestern edge of the state's Northeast Kingdom. This is a place of bedrock ridges, moss-covered granite boulders, boreal forest, and blueberry bushes. It turns out that Kettle Pond is not a kettle pond. Instead it was named either for an eighteenth-century adventurer who lost his kettle or for a nineteenth-century bankrobber who buried a kettle full of money there. Within New England, Vermont may have the distinction of having both the largest single lake and the fewest kettles.

## WISCONSIN

To describe the kettles of the Badger State, one must resort to superlatives. This is where the term "kettle" originated, thanks to T. C. Chamberlain. This is where the nation's longest continuous kettle moraine stretches from Illinois to the border of Upper Peninsula Michigan. This is where the strongest juxtaposition occurs between the dense concentration of kettles in Vilas County and their complete absence in the driftless area to the southwest near the Illinois border. Wisconsin is the home of **Trout Lake,** the focus of a permanent scientific field station operated by the University of Wisconsin since 1924: 85 percent of the shoreline is state owned. More than 2,500 lakes and their related aquatic ecosystems lie within about thirty miles of the station. Trout Lake may be the most well-studied kettle in the nation, explaining why the National Science Foundation selected it as a Long Term Ecological Reserve to monitor how ecosystems will respond to climate change. The University of Wisconsin is the only Big Ten, land-grant university built within a kettle moraine, in this case that of **Lake Mendota,** one of the best long-term case studies of lake pollution.

# Appendix B: Identifying Kettles

To determine if a lake is a kettle or not, I offer several suggestions. A combination of personal, on-the-ground experience and an examination of local maps is usually necessary. For local mapping, a series of state-by-state *Atlas & Gazetteers of the United States*, published by Delorme Inc., is a good resource, being an inexpensive set of paper maps that are easily obtained and show road/city cartography, topography, latitude and longitude, and global positioning coordinates. They also index water bodies and natural areas of interest. A variety of online resources are available, notably Google Earth for aerial imagery (http://earth.google.com), the U.S. Geological Survey for surface materials (http://www.usgs.gov), and the U.S. Natural Resources Conservation Service for soils and drainage (http://soils.usda.gov).

LOCATION: Laurentide kettles occur only north of the ice-sheet limit. The first drawn illustration in this book shows the glacial limit. Atlases of the United States, located in the reference section of most libraries, often show the glacial border as well.

MAP PATTERN: Look for clusters of irregularly shaped lakes. Though the cluster may be elongated, the lakes within it are usually not aligned. If one or more lakes in the cluster has a nearly circular shape, all are likely kettles. Another clue is when roughly circular bays intersect to produce triangular peninsulas within the larger lake.

TOPOGRAPHY: Kettles usually have steep bluffs at least twenty feet high along much of the perimeter. The surrounding terrain is usually either remarkably flat (outwash plain) or highly irregular (ice stagnation). U.S. Geological Survey topographic quadrangle maps usually provide the best information. Their mapping is usually the base for maps of state and local parks and natural areas, having contour lines. If you are using a bathy-metric map, look for a highly irregular lake bottom. Another clue to a kettle lake origin is a lake over eighty feet deep in an area where rock outcrops are not common.

MATERIALS: By definition, kettles are surrounded by collapsed glacial debris and usually dominated by sand and gravel. There is no substitute for looking at road cuts and excavations by yourself to determine the local material. The best map references come from U.S. Geological Survey geology maps (surficial maps if they are available) published at a variety of scales. A lake surrounded by a unit labeled Ice-Contact, Ice-Contact-Stagnation, Outwash, or Kame Terrace is almost certainly a kettle. Alternatively, those surrounded by excessively drained soils, especially coarse sand and gravel, are probably collapsed glacial meltwater deposits; the lakes being kettles. Information officers at federal, state, and local agencies dealing with land and water are great sources of help.

NAMES: The names given to lakes are often clues. Though not definitive, the following words are highly suggestive of a kettle origin: "round," "echo," "sandy," "mud," "bass," "rice," "white," and "kettle."

# Bibliography

Allen, James, Lt. "Journal and Letters" in *Schoolcraft's Expedition to Lake Itasca, Appendix C: Journal and Letters of Lieutenant James Allen.* Philip P. Mason, ed., 163–241. East Lansing: Michigan State University Press, 1993.

Anonymous. *New Guide for Emigrants to the West Containing Sketches of Ohio.* Boston: J. M. Peck, Gould, Kendall, and Lincoln, 1836.

Arvin, Newton. *Longfellow: His Life and Work.* New York: Little, Brown, 1962.

Atwood, Wallace, W. *Physiographic Provinces of North America.* Boston: Ginn, 1940.

Baker, Carlos. *Ernest Hemingway: A Life Story.* New York: Scribner, 1969.

Baraga, Frederic. *A Dictionary of the Ojibway Language.* Minneapolis: Minnesota Historical Society Press, 1992. (Originally *A Dictionary of the Otchipwe Language.* Montreal: Beauchemin & Valois, 1878.)

Birks, Hilary H., M. C. Whiteside, D. M. Stark, and R. C. Bright. "Recent Paleolimnology of Three Lakes in Northwestern Minnesota." *Quaternary Research* 6 (1976): 249–72.

Black, Robert F. *Geology of Ice Age National Scientific Reserve of Wisconsin.* National Park Service Scientific Monograph Series, no. 2. Washington, D.C.: U.S. National Park Service, 1974.

Blanchard, Paula. *Margaret Fuller: From Transcendentalism to Revolution.* New York: Delacorte Press, 1978.

Blegen, Theodore, C. *Minnesota: A History of the State.* Minneapolis: University of Minnesota Press, 1962.

Bolles, Edmund B. *The Ice Finders: How a Poet, a Professor, and a Politician Discovered the Ice Age.* Washington, D.C.: Counterpoint, 1999.

Boothroyd, Jon C., and Les Sirkin. "Quaternary Geology and Landscape Development of Block Island and Adjacent Regions" in *The Ecology of Block Island: Proceedings of the Rhode Island Natural History Survey Conference, October 28, 2000.* Peter Paton, Lisa L. Gould, Peter V. August, and Alexander O. Frost, eds., 13–27. Kingston, RI: Rhode Island Natural History Survey, 2002.

Borst, Raymond R. *The Thoreau Log: A Documentary Life of Henry David Thoreau*. New York: Macmillan, 1992.

Boulton, G. S., and C. D. Clark. "A Highly Mobile Laurentide Ice Sheet Revealed by Satellite Images of Glacial Lineations." *Nature* 346 (1990): 813–17.

Brooks, John L., and Edward S. Deevey Jr. "New England" in *Limnology in North America*. David G. Frey, ed., 117–62. Madison: University of Wisconsin Press, 1963.

Brown, Dee. *Bury My Heart at Wounded Knee*. New York: Holt, Rinehart and Winston, 1970.

Buell, Lawrence. *The Environmental Imagination: Thoreau, Nature Writing, and the Formation of American Culture*. Cambridge, MA: Belknap Press of Harvard University Press, 1995.

Burgis, Mary J., and Pat Morris. *The Natural History of Lakes*. Cambridge, UK: Cambridge University Press, 1987.

Caldwell, D. H., et al. *Surficial Geologic Map of New York* (scale 1:250,000). Albany: New York State Museum, Geological Survey, Map and Chart Series No. 40, 1986–1991. (Note: Five separate sheets: Finger Lakes, 1986, with Ernest H. Muller; Hudson-Mohawk, 1987, with Robert Dineen; and Niagara, 1988, Lower Hudson, 1989, and Adirondack, 1991, with Donald L. Pair.)

Castillo, Ana. "Illinois" in *These United States: Original Essays by Leading American Writers on Their State Within the Union*. John Leonard, ed., 122–31. New York: Nation Books, 2003.

Catton, Bruce. *Michigan: A Bicentennial History*. New York: W.W. Norton, 1976.

Chamberlain, T. C. "On the Extent and Significance of the Wisconsin Kettle Moraine." *Wisconsin Academy of Science, Arts, and Letters Transactions* 4 (1878): 201–34.

Channing, William Ellery. *Thoreau, the Poet-Naturalist: With Memorial Verses*. F. B. Sanborn, ed. Boston: Charles Goodspeed, 1902.

Charlesworth, J. K. *The Quaternary Era with Special Reference to Its Glaciation*, vol. 1. Belfast, Ireland: Edward Arnold, 1957.

Christiansen, Jens Hesselbjerg, and Bruce Hewitson. "Regional Climate Projections" in *Climate Change 2007: The Physical Science Basis. Contribution of Working Group I to the Fourth Assessment Report of the Intergovernmental Panel on Climate Change*. Cambridge, UK: Cambridge University Press, 2007.

Citro, Joseph, A. *Weird New England*. New York: Sterling Publications, 2005.

Clark, Peter U. "Surface Form of the Southern Laurentide Ice Sheet and Its Implications to Ice-Sheet Dynamics." *Geological Society of America Bulletin* 104 (1992): 595–605.

————, and Joseph S. Walder. "Subglacial Drainage, Eskers, and Deforming Beds Beneath the Laurentide and Eurasian Ice Sheets." *Geological Society of America Bulletin* 106 (1994): 304–14.

Clayton, Lee, and Stephen R. Moran. "A Glacial Process-Form Model" in *Glacial Geomorphology, Proceedings of the Fifth Annual Geomorphology Symposia Series.* Donald Coates, ed., 89–119. Binghamton: State University of New York, 1974.

————, with S. R. Moran, J. P. Bluemle, and C. G. Carlson. *Geologic Map of North Dakota* (scale 1:500,000). Reston, VA: U.S. Geological Survey and North Dakota Geological Survey, 1980.

COHMAP Members, "Climatic Changes of the Last 18,000 Years: Observations and Model Simulations." *Science* 241 (1988): 1043–52.

Colman, John A., and Marcus C. Waldron. *Walden Pond, Massachusetts: Environmental Setting and Current Investigations.* U.S. Geological Survey Web site: http://ma.water.usgs.gov/publications/pdf/wal_66.pdf (accessed August 2008).

Conn, Peter. *Literature in America: An Illustrated History.* Cambridge, UK: Cambridge University Press, 1989.

Connors, Judith. "Biography of Walt Whitman" in *Walt Whitman 1819–1892: Criticism and Interpretation.* Harold Bloom, ed., 5–54. Broomall, PA: Chelsea House Publications, 2003.

Crawford, G. W., and David G. Smith. "Palaeoethnobotany in the Northeast" in *People and Plants in Ancient Eastern North America.* Paul E. Minnis, ed., 172–257. Washington, D.C.: Smithsonian Institution Press, 2003.

Cronon, William. *Nature's Metropolis: Chicago and the Great West.* New York: W.W. Norton, 1991.

Current, Richard Nelson. *Wisconsin: A Bicentennial History.* New York: W. W. Norton, 1977.

Davis, Margaret Bryan. "Phytogeography and Palynology of Northeastern United States" in *The Quaternary of the United States.* H. E. Wright Jr. and David G. Frey, eds., 377–401. Princeton: Princeton University Press, 1965.

————. "Holocene Vegetational History of the Eastern United States" in *Late-Quaternary Environments of the United States*, vol. 2, *The Holocene.* H. E. Wright Jr., ed., 166–81. Minneapolis: University of Minnesota Press, 1983.

Deevey, E. S., Jr. "Studies on Connecticut Lake Sediments 1: A Postglacial Climatic Chronology for Southern New England." *American Journal of Science* 237 (1939): 691–724.

Defebaugh, James Elliott. *History of the Lumber Industry of America*, vol. 2. Chicago: The American Lumberman, 1907.

Deloria, Vine, Jr. *God Is Red.* New York: Grosset & Dunlap, 1973.

Dennis, Jerry. *The Living Great Lakes: Searching for the Heart of the Inland Seas.* New York: St. Martin's Press, 2003.

Densmore, Frances. *Chippewa Customs.* Washington, D.C.: U.S. Government Printing Office, Bulletin 86 of Smithsonian Institution Bureau of American Ethnology, 1929. (Reprinted with an introduction by Nina Marchetti Archabal by the Minnesota Historical Society Press, 1979.)

Denton, George H., and Terrence J. Hughes. *The Last Great Ice Sheets.* New York: John Wiley & Sons, 1981.

DeVoto, Bernard (ed.). *The Journals of Lewis and Clark.* Boston: Houghton Mifflin, 1953.

Dingman, S. Lawrence. *Physical Hydrology.* 2nd ed. New York: Prentice Hall, 2002.

Donahue, Brian. *The Great Meadow, Farmers and the Land in Colonial Concord.* New Haven: Yale University Press, 2004.

Dyke, Arthur S., Lynda Dredge, and Jean-Serve Vincent. "Configuration and Dynamics of the Laurentide Ice Sheet During the Late Wisconsin Maximum." *Geographie Physique et Quaternaire* 36 (1982): 5–14.

————, J-S. Vincent, J. T. Andrews, L. A. Dredge, and W. R. Cowan. "The Laurentide Ice Sheet and an Introduction to the Quaternary Geology of the Canadian Shield" in *Quaternary Geology of Canada and Greenland: Geology of North America,* vol. K-1. R. J. Fulton, ed., 178–89. Boulder: Geological Society of America, 1989.

Egerton, Frank N. "The Scientific Contributions of François-Alphonse Forel, the Founder of Limnology." *Aquatic Sciences* 24 (1962): 181–99.

Embleton, Clifford, and C. A. M. King. *Glacial and Periglacial Geomorphology.* London: Edward Arnold, 1969.

Erdich, Louise. *Books and Islands in Ojibwe Country.* Washington, D.C.: National Geographic Society, 2003.

Fagan, Brian M. *The Little Ice Age: How Climate Made History, 1300–1850.* New York: Perseus Books, 2000.

————. *Ancient North America: The Archaeology of a Continent.* New York: Thames & Hudson, 2000.

Farrand, W. R., and D. L. Bell. *Quaternary Geology of Northern Michigan* and *Quaternary Geology of Southern Michigan* (scale 1:500,000). Ann Arbor: State of Michigan Dept. of Natural Resources, Geological Survey Division, 1982.

Fife, Emerson, and Archibald Freeman. *A Book of Old Maps Delineating American History.* New York: Dover, 1969.

Finch, Robert. *The Primal Place.* New York: W.W. Norton, 1983.

Finlayson, C. Max, Rebecca D'Cruz, et al. "Inland Water Systems" in *Current State & Trends Assessment, Millennium Ecosystem Assessment*. http:// www.millenniumassessment.org/documents/document.289.aspx.pdf (accessed August 2008).

Fisher, Daniel C. "Taphonomic Analysis of Late Pleistocene Mastodon Occurrences: Evidence of Butchery by North American Paleo-Indians." *Paleobiology* 10 (1984): 338–57.

Flannery, Tim. *The Eternal Frontier: An Ecological History of North America and Its Peoples*. New York: Grove/Atlantic, 2002.

Flint, Richard Foster. "Glacial Geology" in *Geology, 1888–1938*. 19–41. Boulder: Geological Society of America Fiftieth Anniversary Volume, 1940.

———. *Glacial Map of the United States East of the Rocky Mountains* (scale 1:1,750,000). New York: Geological Society of America, 1959.

Fortey, Richard. *Earth: An Intimate History*. New York: Vintage Books, 2005.

Frazier, Ian. *The Fish's Eye: Essays about Angling and the Outdoors*. New York: Farrar, Straus and Giroux, 2003.

Freese, Ed. L. "Freda Haffner Kettlehole State Preserve" in *Iowa Native Plant Society Newsletter*, August 1996, 3–5.

Frey, David G. (ed.). *Limnology in North America*. Madison: University of Wisconsin Press, 1963.

Fritz, Sherilyn. "Lacustrine Perspectives on Holocene Climate" in *Global Change in the Holocene*. Anson Mackay et al., eds., 227–41. New York: Oxford University Press, 2003.

Gaddis, John Lewis. *The Landscape of History: How Historians Map the Past*. New York: Oxford University Press, 2002.

Gastil, Raymond D. *Cultural Regions of the United States*. Seattle: University of Washington Press, 1975.

Goldthwait, James W. *Surficial Geology of New Hampshire* (scale 1:1,750,000). Concord: New Hampshire State Planning and Development Commission, 1950.

Gray, Henry H. *Quaternary Geologic Map of Indiana* (scale 1:500,000). Indianapolis: Indiana Department of Natural Resources, Geological Survey, Miscellaneous Map 49, 1989.

Grayson, Donald, and David J. Meltzer. "Clovis Hunting and Large Mammal Extinction: A Critical Review of the Evidence." *Journal of World Prehistory* 16 (December 2002): 313–59.

Grimm, Eric C., Pietra Mueller, Jim Clark, James E. Almendinger, Daniel R. Engstrom, Sherilyn Fritz, Emi Ito, and Alisson Smith. "Paleohydrological Significance of Struvite Occurrence in Mid-Holocene Sediments of Kettle Lake, Western North Dakota." *Geological Society of America Abstracts with Programs*, 2005 Annual Meeting, paper no. 105–11.

Grotzinger, John, Thomas H. Jordon, Frank Press, and Raymond Siever. *Understanding Earth*. 5th ed. New York: Freeman, 2007.

Gustavson, T. C. and J. C. Boothroyd. "A Depositional Model for Outwash, Sediment Sources, and Hydrologic Characteristics, Malaspina Glacier Alaska: A Modern Analog of the Southeastern Margin of the Laurentide Ice Sheet." *Geological Society of America Bulletin* 99 (1987): 187–200.

Guthrie, R. Dale. *Frozen Fauna of the Mammoth Steppe: The Story of Blue Babe*. Chicago: University of Chicago Press, 1990.

Hadley, David W., and James H. Pelham (compilers). *Glacial Deposits of Wisconsin: Sand and Gravel Resource Potential* (scale 1:500,000). Madison: Wisconsin Geological and Natural History Survey, Map 10, 1976.

Hagen, Joel B. *An Entangled Bank: The Origins of Ecosystem Ecology*. New Brunswick, NJ: Rutgers University Press, 1992.

Hall, Donald. *Eagle Pond*. Boston: Houghton Mifflin, 2007.

Halprin, Lewis, and Alan Kattelle. *Lake Boon, Massachusetts*. Charleston, SC: Arcadia Publishing (Postcard History Series), 2005.

Harding, Walter, and Carl Bode. *The Correspondence of Henry David Thoreau*. New York: New York University Press (reprinted Westport, CT: Greenwood Press), 1958.

———. *The Days of Henry Thoreau: A Biography*. New York: Dover, 1962.

Harris, Sandra L. and Peter A. Steeves. *Identification of Potential Public Water-Supply Areas of the Cape Cod Aquifer, Massachusetts, Using a Geographic Information System*. Marlborough, MA: U.S. Geological Survey Water-Resources Investigations Report 94-4156, 1994.

Harrison, Jim. "Michigan" in *These United States: Original Essays by Leading American Writers on Their State Within the Union*. John Leonard, ed., 198–205. New York: Nation Books, 2003.

Hawthorne, Nathaniel. "Hobnobbing" in *Walking in America*. Donald Zochert, ed., 59–64. New York: Alfred A. Knopf, 1974. (Reprinted from Simpson, Claude M. [ed]. "The American Notebooks" in *Centenary Edition of the Works of Nathaniel Hawthorne*, vol. 3. William Charvat, Roy Harvey Pearce, and Claude M. Simpson, eds. Columbus: Ohio State University Press, 1973–1974.)

Heat Moon, William Least. *Blue Highways: A Journey into America*. New York: Little, Brown, 1983.

Hemingway, Ernest. *The Complete Short Stories of Ernest Hemingway*. The Finca Vigia edition. New York: Scribner, 1987.

Hobbs, Howard C., and Goebel, Joseph E. *Geologic Map of Minnesota: Quaternary Geology* (scale 1:500,000). Minneapolis: Minnesota Geological Survey, 1982.

Hughes, Robert. *A Jerk at One End: Reflections of a Mediocre Fisherman.* New York: Ballantine Books, 1999.

Hutchinson, G. E. *A Treatise on Limnology,* vol. 1, *Geography, Physics and Chemistry.* New York: John Wiley & Sons, 1957.

———. "The Prospect Before Us" in *Limnology in North America.* David G. Frey, ed., 683–90. Madison: University of Wisconsin Press, 1963.

Illinois State Museum of Natural History. FAUNMAP: An Electronic Database Documenting Later Quaternary Distributions of Mammal Species. http://www.museum.state.il.us/research/faunmap/ (accessed August 2008).

Itasca State Park. *The Itasca Guidebook: A Complete Guide to the Natural and Cultural History and Recreational Resources of Itasca State Park.* Updated ed. (Available at Itasca State Park Gift Shop/Bookstore.)

Jackson, Julia A. *Glossary of Geology.* 4th ed. Washington, D.C.: American Geological Institute, 2005.

Jacobson, George L., Thompson Webb, and Eric C. Grimm. "Changing Vegetation Patterns of Eastern North America During the Past 18,000 Years: Inferences from Overlapping Distributions of Selected Pollen Types" in *North America and Adjacent Oceans During the Last Deglaciation. Decade of North America,* vol. K-3. William Ruddiman, ed. Boulder: Geological Society of America, 1987.

Jensen, O. P., B. J. Benson, J. J. Magnuson, V. M. Card, M. N. Futter, P. A. Soroanno, and K. M. Steward. "Spatial Analysis and Ice Phenology Trends Across the Laurentian Great Lakes Region During a Recent Warming Period." *Limnology & Oceanography* 52 (2007), 2013–26.

Johansen, Bruce E. *The Dirty Dozen: Toxic Chemicals and the Earth's Future.* Westport, CT: Praeger Publishers, Greenwood Publishing Group, 2003.

Johnson, Charles W. *Bogs of the Northeast.* Hanover, NH: University Press of New England, 1985.

Keillor, Garrison. *Lake Wobegon Days.* New York: Viking Penguin, 1985.

———. *In Search of Lake Wobegon.* New York: Viking Penguin, 2001.

———. *Lake Wobegon Summer, 1956.* New York: Viking Penguin, 2001.

Kenton, Edna (ed.). *The Jesuit Relations and Allied Documents: Travel and Explorations of the Jesuit Missionaries in North America (1610–1791), with an Introduction by Reuben Goldthwaite.* New York: Albert & Charles Boni, 1925.

Kipfer, Barbara Ann. *Encyclopedic Dictionary of Archaeology.* New York: Kluwer Academic, 2000.

Kopper, Philip. *The Smithsonian Book of North American Indians: Before the Coming of the Europeans.* Washington, D.C.: Smithsonian Books, 1986.

Koteff, Carl. "Glacial Lakes near Concord, Massachusetts" in *U. S. Geologi-

*cal Survey, Professional Paper 475-C.* Washington, D.C.: U.S. Government Printing Office, 1963.

LaDuke, Winona. *Recovering the Sacred: The Power of Naming and Claiming.* Cambridge, MA: South End Press, 2005.

———. *All Our Relations: Native Struggles for Land and Life.* Cambridge, MA: South End Press, 1999.

Laskin, David. *The Children's Blizzard.* New York: HarperCollins, 2004.

Lass, William E. *Minnesota: A Bicentennial History.* New York: W.W. Norton, 1977.

Leakey, Richard, and Roger Lewin. *People of the Lake: Mankind and Its Beginnings.* New York: Doubleday, 1978.

Lemke, R. W., W. M. Laird, M. J. Tipton, and R. M. Lindvall. "Quaternary Geology of Northern Great Plains" in *The Quaternary of the United States.* H. E. Wright Jr. and David G. Frey, eds., 15–27. Princeton: Princeton University Press, 1965.

Leopold, Aldo. *A Sand County Almanac, and Sketches Here and There.* New York: Oxford University Press, 1949.

Lewis, Sinclair. *Main Street* (with an introduction and notes by Brooke Allen). New York: Barnes & Noble Classics, 2003 (originally published 1920).

Licciardi, Joseph M., James T. Teller, and Peter U. Clark. "Freshwater Routing by the Laurentide Ice Sheet During the Last Deglaciation" in *Mechanisms of Global Climate Change at Millennial Time Scales.* Peter U. Clark, Robert S. Webb, and Lloyd D. Keigwin, eds., 177–201. Washington, D.C.: American Geophysical Union Monograph 112, 1999.

Lineback, Jerry A. (compiler). *Quaternary Deposits of Illinois* (scale 1:500,000). Urbana: Illinois State Geological Survey, 1979.

Lingeman, Richard. *Sinclair Lewis: Rebel from Main Street.* New York: Random House, 2002.

Löfgren, Orvar. *On Holiday: A History of Vacationing.* Berkeley: University of California Press, 1999.

Long, Robert Emmet. *James Fenimore Cooper.* New York: Continuum, 1990.

Longfellow, Henry Wadsworth. *Hiawatha* in *The Complete Poetical Works of Longfellow.* 113–63. Cambridge, MA: Riverside Press, Houghton Mifflin, 1886 (reprinted in 1920).

Louv, Richard. *Last Child in the Woods: Saving Our Children from Nature-Deficit Disorder.* Chapel Hill, NC: Algonquin Books, 2005.

Lynd, Robert S., and Helen Merrell Lynd. *Middletown: A Study in American Culture.* New York: Harcourt, Brace & World, Inc., 1929.

Lyon, Thomas J. *This Incomparable Land: A Guide to American Nature Writing.* Minneapolis: Milkwood Editions (nonprofit), 2001.

Maizels, J. "Sediments and Landforms of Modern Proglacial Terrestrial Environments" in *Glacial Environments,* vol. 1, *Modern Glacial Environments: Processes, Dynamics and Sediments.* John Menzies, ed., 365–416. Oxford, UK: Butterworth-Heinemann, 1995.

Mann, Charles C. *1491: New Revelations of the Americas Before Columbus.* New York: Alfred A. Knopf, 2005.

Marquart, Debra. *The Horizontal World: Growing Up Wild in the Middle of Nowhere.* New York: Perseus Books, 2006.

Martin, Paul S., and Richard G. Klein (eds). *Quaternary Extinctions.* Tucson: University of Arizona Press, 1984.

Mason, Philip P. (ed.). *Schoolcraft's Expedition to Lake Itasca: The Discovery of the Source of the Mississippi.* East Lansing: Michigan State University Press, 1993.

Masterson, J. P., and John W. Portnoy. *Potential Changes in Ground-Water Flow and Their Effects on the Ecology and Water Resources of the Cape Cod National Seashore, Massachusetts.* Washington, D.C.: U.S. Geological Survey General Information Product 13, 2005.

Matheus, Paul E. *Paleoecology and Ecomorphology of the Giant Short-Faced Bear in Eastern Beringia.* Fairbanks: University of Alaska Ph.D. dissertation, 1997.

Maynard, W. Barksdale. *Walden Pond, a History.* New York: Oxford University Press, 2004.

McAndrews, John H. "Pollen Analysis and Vegetational History of the Itasca Region, Minnesota" in *Quaternary Paleoecology.* E. J. Cushing and H. E. Wright Jr., eds., 219–36. New Haven: Yale University Press, 1967.

McMichael, George (ed.). *Anthology of American Literature,* vol. 1, *Colonial Through Romantic.* New York: Macmillan, 1989.

Meehl, G. A., et al. "2007 Global Climate Projections" in *Climate Change 2007: The Physical Science Basis. Contribution of Working Group I to the Fourth Assessment Report of the Intergovernmental Panel on Climate Change.* Susan Solomon et al., eds., 749–51. Cambridge, UK: Oxford University Press, 2007.

Merk, Frederick. *History of the Westward Movement.* New York: Alfred A. Knopf, 1978.

Mickelson, D. M., Lee Clayton, D. S. Fullerton, and H. W. Borns Jr. "The Late Wisconsin Glacial Record of the Laurentide Ice Sheet in the United States" in *Late-Quaternary Environments of the United States,* vol. 1, *The Late Pleistocene.* Stephen C. Porter, ed., 3–37. Minneapolis: University of Minnesota Press, 1983.

Minnesota State Department of Natural Resources. "Frequently Asked

Questions / Minnesota Facts / Water." http://www.dnr.state.mn.us/faq/ mnfacts/water.html (accessed August 2008).

Mitchell, John Hanson. *Ceremonial Time: Fifteen Thousand Years on One Square Mile.* Reading, MA: Addison Wesley, 1984.

Mitsch, William J., and James G. Gosselink. *Wetlands.* 3rd ed. New York: John Wiley & Sons, 2000.

Moberg, Vilhelm. *The Last Letter Home* (originally under the title *Nybyggarna and Sist Bretet Till Syvrige,* translated from Swedish by Gustaf Lannestock). New York: Simon & Schuster, 1961.

Morrison, Samuel Eliot. *The European Discovery of America: The Northern Voyages.* New York: Oxford University Press, 1971.

Mosher, Howard Frank. *North Country: A Personal Journey Through the Borderland.* Boston: Houghton Mifflin, 1977.

Muir, Diana. *Reflections in Bullough's Pond: Economy and Ecosystem in New England.* Hanover, NH: University Press of New England, 2000.

Muir, John. "The Story of My Boyhood and Youth" in *Nature Writings: The Library of America.* William Cronon, ed., 33–64. New York: Viking Penguin (Literary Classics of the United States), 1997.

National Assessment Synthesis Team (U.S. Global Change Research Program). *Climate Change Impacts on the United States: The Potential Consequences of Climate Variability and Change.* Washington, D.C.: U.S. Government Printing Office, 2001.

National Geographic Society. *Historical Atlas of the United States.* Centennial ed. Washington, D.C.: National Geographic Society, 1988.

Nester, Peter, and Larry Brown. "Three Dimensional Geometry of a Late Pleistocene Basin, Hyde Park, New York, Integrating Field Observation Techniques and Ground Penetrating Radar" (abstract). Meeting held Lunenburg, Nova Scotia, March 28, 2003, by the Geological Society of America Northeastern Section.

North, Sterling. *Rascal: A Memoir of a Better Era.* New York: Dutton, 1963.

Oldale, Robert N. *Cape Cod, Martha's Vineyard & Nantucket: The Geologic Story* (revised and updated). Yarmouthport, MA: On Cape Publications, 2001.

———, and C. J. O'Hara. "Glaciotectonic Origin of the Massachusetts Coastal End Moraines and a Fluctuating Late Wisconsinan Ice Margin." *Geological Society of America Bulletin* 95 (1984): 61–74.

Olson, Sigurd F. "Northern Lights" in *The Norton Book of Nature Writing.* College ed. Robert Finch and John Elder, eds., 433–35. New York: W.W. Norton, 2002. (Originally published in Sigurd F. Olson, *The Singing Wilderness,* New York: Knopf, 1956.)

————. *Runes of the North*. New York: Alfred A. Knopf, 1963.

————. *Of Time and Place*. New York: Alfred A. Knopf, 1982.

Outwater, Alice. *Water, a Natural History*. New York: Basic Books, 1996.

Peltier, W. Richard. "Ice Age Paleotopography." *Science* 265 (July 1994): 195–201.

Pielou, E. C. *After the Ice Age: The Return of Life to Glaciated North America*. Chicago: University of Chicago Press, 1991.

————. *Fresh Water*. Chicago: University of Chicago Press, 1998.

Poreba, David L. *On the Road Histories: Michigan*. Northampton, MA: Interlink Books, 2006.

Portnoy, J. W., M. G. Winkler, P. R. Sanford, and C. N. Farris. *Kettle Pond Data Atlas: Paleoecology and Modern Water Quality* (119 page report). Wellfleet, MA: U.S. Dept. of the Interior, National Park Service, Cape Cod National Seashore, 2001.

Prest, V. K. "Quaternary Geology" in *Geology and Economic Minerals of Canada*. R. J. W. Douglas, ed., 675–764. Ottawa: Canadian Geological Survey, 1970.

Radway-Stone, Janet, John P. Schafer, Elizabeth Haley-London, Mary L. DiGiacomo-Cohen, Ralph S. Lewis, and Woodrow B. Thompson. *Quaternary Geologic Map of Connecticut and Long Island Sound Basin* (scale 1:125,000). Reston, VA: U.S. Geological Survey, Map MIM 2784, 2005.

Reed, John C., Jr., John O. Wheeler, and Brian E. Tucholke. *Geologic Map of North America* (scale 1:5,000,000). Boulder: Geological Society of America, 2005.

Richardson, Robert D., Jr. *Emerson: The Mind on Fire*. Berkeley: University of California Press, 1995.

Ritchie, William A. *The Lamoka Lake Site: The Type Station of the Archaic Algonkin Period in New York: Researches and Transactions*, vol. 7, no. 1. Rochester: New York State Archaeological Association, Lewis H. Morgan Chapter, 1932.

Robinson, Elwyn B. *History of North Dakota*. Lincoln: University of Nebraska Press, 1966.

Robinson, Michelle. *The Massachusetts Lake and Pond Guide: Protection Through Education*. Boston: Massachusetts Department of Conservation and Recreation, Office of Water Resources, Lakes and Ponds Program, 2004.

Rølvaag, Ole Edvart. *Giants in the Earth: A Saga of the Prairie*. New York: Harper & Brothers, 1927. (Originally published in Norwegian as *I de Dage*; by Lincoln Colcord and the author.)

Rostland, Erhard. *Freshwater Fish and Fishing in Native North America*. California Publications in Geography, vol. 9. Berkeley: University of California Press, 1952.

Ruess, Henry S., Gilbert Tanner, Philip Dinsmoor, and Robert Hellman. *On the Trail of the Ice Age: A Hiker's and Biker's Guide to Wisconsin's Ice Age National Scientific Reserve and Trail.* Milwaukee: Milwaukee Public Museum, 1976.

Sandburg, Carl. *The Complete Poems of Carl Sandburg* (revised and expanded edition with an introduction by Archibald MacLeish). New York: Harcourt Brace Jovanovich, 1970.

Schoolcraft, Henry Rowe. *Schoolcraft's Expedition to Lake Itasca: The Discovery of the Source of the Mississippi* (originally published as *Schoolcraft's Narrative of an Expedition Through the Upper Mississippi to Itasca Lake*). New York: Harper & Brothers. Reissued in 1855 as *Schoolcraft's Summary and Narrative of an Exploratory Expedition to the Sources of the Mississippi River in 1820: Resumed and Completed by the Discovery of Its Origin in Itasca Lake in 1832.* Philip Mason, ed. East Lansing: Michigan State University Press, 1993.

————. *The Myth of Hiawatha, and Other Oral Legends, Mythologic and Allegoric, of the North American Indians.* Philadelphia: J. B. Lippincott & Company, 1856. (Reprinted with a supplemental introduction by Larry B. Massie, 1884.)

Setterdahl, Lilly. *Minnesota Swedes: The Emigration from Torlle Ljungby to Goodhue County, 1855–1912.* East Moline, IL: American Friends of the Emigrant Institute of Sweden, Inc., 1996.

Sevon, W. D., and D. D. Braun. *Glacial Deposits of Pennsylvania* (scale not reported). Harrisburg, PA: Conservation and Natural Resources, Bureau of Topographic and Geologic Survey, Map 2200-MP-DCNR3027, 1997.

Sevon, W. D., and Gary M. Fleeger. "Pennsylvania and the Ice Age." *Educational Series* 6. Philadelphia: Pennsylvania Geological Survey, 1999.

Shay, C. T. *The Itasca Bison Kill Site: An Ecological Analysis.* St. Paul: Minnesota Historical Society (Minnesota Prehistoric Archaeology Series), 1971.

Shephard, Lansing. *The Northern Plains: Minnesota, North Dakota, and South Dakota.* Washington, D.C.: Smithsonian Books (The Smithsonian Guides to Natural America), 1996.

Shoemaker, Nancy. *A Strange Likeness: Becoming Red and White in Eighteenth-Century North America.* New York: Oxford University Press, 2004.

Smith, John. "A Description of New England" in *Anthology of American Literature*, vol. 1, *Colonial Through Romantic.* George McMichael, ed., 24–32. New York: Macmillan, 1989.

Soller, David R. *Map Showing the Thickness and Character of Quaternary Sediments in the Glaciated United States East of the Rocky Mountains* (scale 1:3,500,000). U.S. Geological Survey Miscellaneous Investigations Series Map I-1970-E, 2001.

Spencer, Robert F., et al. *The Native Americans.* New York: Harper & Row, 1977.

Stanford, Scott D., and Ron W. Witte. *Surficial Geology of New Jersey* (scale 1:100,000). New Jersey Department of Environmental Protection, New Jersey Geological Survey, 2006.

Steinbeck, John. *Travels with Charley.* New York: Viking Press, 1962.

Stoltman, James B., and David A. Baerreis. "The Evolution of Human Ecosystems in the Eastern United States" in *Late-Quaternary Environments of the United States,* vol. 2, *The Holocene.* H. E. Wright Jr., ed., 252–68. Minneapolis: University of Minnesota Press, 1983.

Swanson, G. A., T. C. Winter, V. A. Adomatis, and J. W. LaBaugh. *Chemical Characteristics of Prairie Lakes in South-Central North Dakota: Their Potential for Impacting Fish and Wildlife.* Washington, D.C.: U.S. Fish and Wildlife Service Technical Report 18, 1988.

Tankersley, Kenneth. *In Search of Ice Age Americans.* Layton, UT: Gibbs Smith, 2002.

Tarr, Ralph Stockman, and Lawrence Martin. *Alaskan Glacier Studies of the National Geographic Society in the Yakutat Bay, Prince William Sound and Lower Copper River Regions. Based Upon the Field Work in 1909, 1910, 1911 and 1913 by National Geographic Society Expeditions.* Washington, D.C.: National Geographic Society, 1914.

Taylor-Wilkie, Doreen. *Insight Guide to Sweden.* Maspeth, NY: Langenscheidt Publishers, Inc., 1999.

Thomas, David Hurst, Jay Miller, Richard White, Peter Nabokov, and Phillip J. Deloria. *The Native Americans: An Illustrated History.* Atlanta: Turner Publishing, Inc., 1993.

Thomas, Joseph D. (ed). *Cranberry Harvest: A History of Cranberry Growing in Massachusetts.* New Bedford, MA: Spinner Publications, Inc., 1990.

Thompson, Carol, Jean Prior, and Deborah Quade. *Quaternary Landforms and Hydrogeology of Fens in Clay and Dickinson Counties or Mucking about in Northwest Iowa.* Des Moines: Geological Society of Iowa Guidebook 44, 1991.

Thompson, Gerald, Jennifer Coldrey, and George Bernard. *The Pond.* Cambridge, MA: MIT Press, 1984.

Thompson, Woodrow B., and Harold W. Borns Jr. (eds.). *Surficial Geologic*

*Map of Maine* (scale 1:500,000). Augusta: Maine Department of Conservation, Geological Survey, 1985.

Thoreau, Henry David. *Cape Cod* (with an introduction by Joseph Wood Krutch and illustrations by R. J. Holden; originally published in 1865). New York: The Heritage Press, 1968.

———. *Walden* and *Resistance to Civil Government, Authoritative Texts, Thoreau's Journal, Reviews and Essays in Criticism.* 2nd ed. William Rossi, ed. (*Walden* originally published as *Walden; or Life in the Woods.* Boston: Ticknor and Fields, 1854.) New York: W.W. Norton, 1962.

Thorson, Robert M., and Gregory Brick. "Stratified Drift as Historical Baggage." *GSA Today* (Geological Society of America) 5 (1995): 174–76.

Thorson, Robert M., and Robert S. Webb. "Postglacial History of a Cedar Swamp in Southeastern Connecticut." *Journal of Paleolimnology* 6 (1991): 17–35.

Torrey, Bradford, and Francis H. Allen (eds.). *The Journal of Henry David Thoreau,* vol. 14, *August 1, 1860–November 3, 1861.* Layton, UT: Peregrine Smith Books, 1984.

Totten, Stanley M., and George W. White. "Glacial Geology and the North American Craton: Significant Concepts and Contributions of the Nineteenth Century" in *Geologists and Ideas: A History of North American Geology,* GSA centennial special vol. 1. Ellen T. Drake and William M. Jordan, eds., 125–41. Boulder: Geological Society of America, 1985.

Trachtenberg, Alan. *Shades of Hiawatha: Staging Indians, Making Americans: 1880–1930.* New York: Hill and Wang, 2005.

U.S. Department of Agriculture (Forest Service). Hubbard Brook Ecosystem Study. http://www.hubbardbrook.org/overview/site_description.htm (accessed October 2008).

U.S. Environmental Protection Agency. *National Lakes Assessment.* http://www.epa.gov/owow/lakes/lakessurvey/ (accessed August 2008).

U.S. Geological Survey. National Water-Quality Assessment (NAWQA) Program. http://water.usgs.gov/nawqa/ (accessed August 2008).

———. Northern Prairie Wildlife Research Center. http://www.npwrc.usgs .gov/resource/wetlands/basinwet/chap2a.htm (accessed August 2008).

Waters, Michael R., and Thomas W. Stafford Jr. "Redefining the Age of Clovis: Implications for the Peopling of the Americas." *Science* 315 (Feb. 23, 2007): 1122–26.

Watson, R. T., et al. (eds.). *Climate Change 2001: Synthesis Report. A Contribution of Working Groups I, II, and III to the Third Assessment Report of the Intergovernmental Panel on Climate Change.* Cambridge, UK, and New York: Cambridge University Press, 2001.

Webb, Thompson, III, Patrick J. Bartlein, Sandy P. Harrison, and Katherine

H. Anderson. "Vegetation, Lake Levels, and Climate in Eastern North America for the Past 18,000 Years" in *Global Climates Since the Last Glacial Maximum.* H. E. Wright Jr. et al., eds., 415–67. Minneapolis: University of Minnesota Press, 1993.

Wetzel, Robert G. *Limnology.* 2nd ed. New York: Saunders College Publishing, 1983.

———. *Limnology: Lake and River Ecosystems.* 3rd ed. New York: Academic Press, 2001.

White, E. B. "Once More to the Lake" in *Best Essays of the Century.* Joyce Carol Oates, ed., 179–85. Boston: Houghton Mifflin, 2000.

———. "Walden—1954" in *Walden* and *Resistance to Civil Government: Authoritative Texts, Thoreau's Journal, Reviews and Essays in Criticism.* 2nd ed. William Rossi, ed., 359–66. New York: W.W. Norton, 1992.

White, William A. "Deep Erosion by Continental Ice Sheets." *Geological Society of America Bulletin* 83 (1972): 1037–56.

Whitman, Walt. *Leaves of Grass.* Inclusive ed. Emory Holloway, ed. Garden City, NY: Doubleday, 1942. (Originally self-published in 1855.)

Whittemann, Betsey, and Nancy Webster. *Water Escapes in the Northeast: Great Waterside Vacation Spots.* West Hartford, CT: Wood Pond Press, 1987.

Whittlesley, Charles. "Notes on the Drift and Alluvium of Ohio and the West." *American Journal of Science* 5 (1848): 205–17.

Wineapple, Brenda. *Hawthorne: A Life.* New York: Alfred A. Knopf, 2003.

Winkler, Marjorie Green. "Changes at Walden Pond During the Last 600 Years: Microfossil Analyses of Walden Pond Sediments" in *Thoreau's World and Ours: A Natural Legacy.* Edmund A. Schofield and Robert C. Baron, eds., 199–211. Golden, CO: North American Press, 1993.

Winslow, Edward (with William Bradford). *Mourts Relation: A Journal of the Pilgrims at Plymouth, 1622* (with an introduction by Dwight B. Heath). Bedford, MA: Applewood Books, 1963.

Winter, T. C., and M. K. Woo. "Hydrology of Lakes and Wetlands" in *The Geology of North America*, vol. 1., *Surface Water Hydrology.* M. G. Wolman and H. C. Riggs, eds., 159–87. Boulder: Geological Society of America, 1990.

Wright, H. E., and Robert V. Ruhe. "Glaciation of Minnesota and Iowa" in *The Quaternary of the United States.* H. E. Wright and David G. Frey, eds., 29–41. Princeton: Princeton University Press, 1965.

Zochert, Donald. *Walking in America.* New York: Alfred A. Knopf, 1974.

Zumberge, James H. "The Lakes of Minnesota: Their Origin and Classification." *Minnesota Geological Survey Bulletin* 35. Minneapolis: University of Minnesota Press, 1952.

# Notes

## INTRODUCTION

1. Winslow, *Mourts Relation*, 20. Subsequent quotes about Pilgrims, ibid., 22, 23.
2. *Oxford English Dictionary*, 1971 compact ed., s.v. "kettle."
3. The authoritative reference comes from Jackson, *Glossary of Geology*, here slightly abbreviated. "Kettle: A steep-sided, usually basin- or bowl-shaped hole or depression, commonly without surface drainage, in glacial drift deposits . . . often containing a lake or swamp; formed by the melting of a large, detached block of stagnant ice . . . that had been wholly or partly buried in the glacial drift." This usage is similar to that of Wetzel (*Limnology*), which follows Hutchinson (*Treatise on Limnology*) and his precedents back to William Morris Davis. Other terms have been attached to the phenomenon, most notably "ice-block" lake (Zumberge, "Lakes of Minnesota"). Specifically, the term "kettle" refers only to the depression. This book uses the term "kettle lake" broadly to include all "ice-block" meltdown depressions, regardless of size, including those dammed by kettle moraines because the materials and mechanisms are similar.
4. Phrase from Finch, *Primal Place*, 109. The usage is correct because "berg" translates as "mountain," rather than "floe." Ice "block" is the standard term for a detached mass of terrestrial ice, regardless of whether it is slab shaped or not.
5. According to Carolyn Freeman-Travers, the water source was identified as Dyer's Swamp in the village of East Harbor in Freeman's *History of Cape Cod* (1801). The pond where they later camped was the one from which Pond Village in Truro was named, based on Henry M. Dexter's 1865 edition of *Mourts Relation*.
6. Oldale, *Cape Cod*, 75–77.
7. Walden Pond, in Concord, Massachusetts, is a perfect example in terms of size and with a shore "irregular enough not to be monotonous" (Thoreau, *Walden*, 125).

8. Hutchinson (*Treatise on Limnology*) provided the first definitive attempt to classify all lakes. Of his seventy-six categories, twenty are related to glaciation. Within North America, lakes carved out of bedrock are the most common because they dominate the Canadian interior. If only the United States is considered, kettle lakes are much more common. I know of no accurate count of the number of kettle lakes, though I estimate there are fifty thousand at a minimum. Many small lakes in New England (partial kettles) are transitional, between a rock-carved and an ice-block origin.

9. Fisher, "Taphonomic Analysis."

10. The phrase is also the title of a book by historian John Lewis Gaddis (2002), who argues for the value of thematic history.

11. In sequence, statements in previous paragraph about: ethnicity from Gastil, *Culture Regions of the United States*, 9; architecture from National Geographic Society, *Historical Atlas*, 42–43. In this paragraph: immigration from Moberg, *Last Letter Home*, 48 and Laskin, *Children's Blizzard*, 27–28; population from National Geographic Society, *Historical Atlas*, 48. This source shows four islands of very low population density in 1890 that correlate with four of the densest concentrations of kettle lake terrain in northern Minnesota, northwestern and northeastern Wisconsin, and the northern part of Michigan's southern peninsula (Soller, *Quaternary Sediments*).

12. Channing, *Thoreau, the Poet-Naturalist*, 15, 115–16. Thoreau scholar Lawrence Buell (*Environmental Imagination*, 525, note 2) writes, "I find it entirely plausible that Thoreau's environmental precocity was facilitated by the transfer of passion for a partner too close to oneself to the safer surrogate of the nonhuman otherworld."

13. Thoreau, *Walden*, 118.

14. The phrase "better things through chemistry" is from Outwater, *Water*, 150. Following quotes from Maynard, *Walden Pond*, 267 and 13.

15. Keillor, *Lake Wobegon Days*, 1, 8.

16. The evidence comes from an unusual phosphate mineral preserved in its guano-rich sediment (Grimm et al., "Paleohydrological Significance of Struvite."

17. This subdivision is based broadly on discrete patterns in the concentrations of sand and gravel (Soller, *Quaternary Sediments*,) generally where it is more than fifty feet thick.

18. Within hilly, rocky terrain, ice blocks are typically stranded in bedrock basins, but the bodies of water within them are dammed in by glacial deposits, giving such lakes a composite origin and hydrologic properties.

19. The southern prong of Ontario would be identified as a fourth northern heartland state were it not in Canada. Geologically, it resembles Michigan, being a former island of glacial drift surrounded on all sides by lobes of ice.
20. Hagen, *Entangled Bank*.
21. Finlayson and D'Cruz, "Inland Water Systems," provide a global overview of the problems associated with small inland lakes, approximately half of which have disappeared in the twentieth century. The main issues are water supply, pollution, and undesirable ecological changes.
22. Louv, *Last Child*, 19.

## 1. ICE-SHEET INVASION

1. There is so little melting in the bitterly cold climate of the South Pole that the main control on glacier mass is the amount of heat that can translate into precipitation of snow. Warm air can hold more moisture than cold, hence delivering more snow to the interior of an ice sheet, provided all this takes place below the freezing point.
2. Terminology follows the conventions of Dyke et al., "Laurentide Ice Sheet," 184. The term Laurentide Ice Sheet is restricted to the last major ice advance in North America only, and refers to the entire coalesced system, which, in map view, was centered over Hudson Bay.
3. For the purposes of hemispheric paleoclimate, the Laurentide can be considered as a single entity (COHMAP Members, "Climatic Changes"; Peltier, "Ice Age Paleotopography") with multiple domes (Dyke et al., "Configuration and Dynamics," 6) and generally higher toward Hudson Bay, where gravity data indicate the ice mass was most concentrated. I do not suggest that the ice sheet was a single, symmetrical, dome-shaped ice mass as do Denton and Hughes, *Last Great Ice Sheets*, fig. 6-5. Controversy remains. The ice sheet was about two miles thick.
4. Richard Fortey's *Earth: An Intimate History* provides an accessible review. Grotzinger et al., *Understanding Earth*, provide a good textbook treatment.
5. Refer to: Allen, "Journal and Letters," 230, for boulder occurrence; Poreba, *On the Road*, 10, for mining culture; and Morrison, *European Discovery*, 410–24, for ethnography.
6. Technically, they are called banded iron formations (BIF). Dark beds can also include the iron oxide magnetite and shale. Flint, as used here, is synonymous with gray chert. The banded ore comes from the buildup

of iron salts in the ocean, followed by their abrupt precipitation as oxides after an infusion of dissolved oxygen. Only later did Earth develop an oxygenated atmosphere.

7. Quote from Heat Moon, *Blue Highways*, 283.

8. Geography and chronology from Boulton and Clark, "Highly Mobile Laurentide."

9. Flow paths for ice at the maximum were modified by later ones, making resolution of ice sources challenging. Prest, "Quaternary Geology," fig. XII-15. Denton and Hughes, *Last Great Ice Sheets*, fig. 6-1, suggest a single-domed radial model. Dyke et al., "Configuration and Dynamics," and Dyke et al., "Laurentide Ice Sheet," review the controversies regarding the multidome interpretation and ice sources.

10. Charlesworth, *Quaternary Era*, 404.

11. Ibid., 376.

12. Although rock is the ultimate source, much of the till associated with kettles in the glaciated fringe was derived by reworking previous tills laid down during at least three previous major episodes predating the Laurentide (Mickelson et al., "Late Wisconsin Glacial Record"). These earlier episodes are named Illinoisan, Kansan, and Nebraskan. A few filled kettles remain from the Illinoisan advance, identified only by coring. The undulating surface above lodgment till is often erroneously referred to as ground moraine. Till plain is more accurate. Being dense and impermeable, low spots on a till plain often contain small ponds, which can easily be confused with kettles from the air.

13. Mickelson et al., "Late Wisconsin Glacial Record," and Licciardi et al., "Freshwater Routing," provide regional summaries of the geology and chronology for different sectors of the ice sheet at the continental scale. Oldale and O'Hara, "Glaciotectonic Origin," explain the general mechanism of moraine formation in New England. Clark, "Surface Form," reviews the glaciology and mass flux of Great Lakes lobes.

14. Ice melts at 32 degrees Fahrenheit (0 degrees Celsius). Depending on its composition, crustal rock melts at 1,200–2,000 degrees Fahrenheit (650–1,100 degrees Celsius). Normal rock is at least ten times stronger than ice under uniaxial compression. Age is not a criterion for rock because some basalt is modern and some ice is millions of years old.

15. If the base of the glacier is considered a geological fault, then the till produced in the fault zone is equivalent to breccia and mylonite, coarse and fine types of rock produced by dynamic motion of brittle faults.

16. The geological term for this type of glide-plane fault is a décollemont. Being driven by collapse, it would also be referred to as a gravity

detachment or sole thrust. Such faults commonly occur within the crust near the base of growing mountain ranges. There are true salt glaciers in the driest parts of Iran, meaning that the concept of glacier need not be restricted to ice, another reason to think of glacier ice as a rock.

17. Called regelation, this is one of the most important mechanisms for eroding bedrock and incorporating debris in the ice. The extra-high pressure on the up-glacier side of bumps and the extra-low pressure on the down-glacier side causes the ice to melt, flow around the obstacle as a thin film of water, then refreeze onto the moving ice, tearing blocks forward in the process. At a much larger scale, there are basal transition zones where the glacier is freezing onto its bed, even as it moves forward. These are often zones of intense erosion.

18. This particle size is referred to as the terminal grade, and is strongly controlled by the source rock being crushed. Early in the process of wet crushing, the larger particles are more readily crushed owing to greater exposure, whereas smaller ones "hide" between average-sized ones or are rinsed away. When the residue becomes dominated by sand and silt, pebbles move within it, becoming rounder and smoother.

19. Technically, sand is any granular material smaller than 2 millimeters (slightly more than 1/16 inch) in intermediate diameter and coarser than 0.063 millimeters (grit). Medium sand approximates the texture of granulated sugar and salt. Boulders are any particle larger than 256 millimeters (about 10 inches).

## 2. THE BIRTH OF KETTLE LAKES AND PONDS

1. Some scientists argue that there was indeed a single dome over Hudson Bay; others argue for a series of domes that shifted through time. Regardless, the center of the crustal depression approximates the center of the mass of the ice sheet over Hudson Bay (Peltier, "Ice Age Paleotopography," 200).

2. Boulton and Clark, "Highly Mobile Laurentide."

3. The archipelago consists of Nantucket, Cape Cod (it's actually an island), Martha's Vineyard, the Elizabeth Islands, Fishers Island, Block Island, Long Island, Staten Island, and others.

4. Gustavson and Boothroyd, "Depositional Model."

5. A review of this history of discovery and geological interpretations is provided by Tarr and Martin, Alaskan *Glacier Studies*, 41–49.

6. The deglacial climate and vegetation were not analogous. Today the vegetation around the Gulf of Alaska is a dense closed-canopy conifer

forest, and the climate is fogbound, fairly warm in summer, and with little snow. In contrast, the southern edge of the Laurentide Ice Sheet during kettle formation was a much drier, windswept tundra mixed with spruce, and with snow on the ground for most of the year (Boothroyd and Sirkin, "Landscape Development of Block Island"). The physical processes were similar, though the rates were far slower in New England.

7. Based on unpublished field work in 1977 reported to the U.S. Geological Survey, the National Park Service, and the National Geographic Society.

8. The pattern and chronology of retreat control the outcome. If the ice supply remained vigorous, the marginal surface profile stayed fairly steep, minimizing stagnation. More commonly, the ice supply weakened, flattening the local surface profile, accentuating the formation of lobes, and maximizing the opportunity for stranding blocks. A fast retreat under warm conditions yielded a high rate of meltwater sediment production but not enough time for blocks to be buried or embanked. Slow retreat under cold conditions allowed plenty of time for burial, but produced less sediment. The ideal conditions for kettle formation occurred when the zone of active ice stepped backward a mile or so, then remained in place to feed copious sediment outward into the stagnant fringe.

9. Rarely, small ice blocks can be floated in swollen meltwater rivers, giving rise to very small kettles (Maizels, "Modern Proglacial," 379).

10. Quoted from Black, *Scientific Reserve of Wisconsin*, 31. Subsequent quotes are from: Schoolcraft, *Expedition to Lake Itasca*, 48; Moberg, *Last Letter*, 16.

11. Also known as Lake Chaubunagungamaug or Lake Webster, Brooks and Deevey, "New England," 118–19, described this as an irregular kettle. The name roughly translates as "you fish on your side and I fish on mine." Thoreau, *Cape Cod*, 106; Beston quote from Zochert, *Walking in America*, 265–66; Ritchie, *Lamoka Lake*, 80.

12. Tills can also be deposited above bedrock and older glacial deposits. Sometimes, house-sized blocks of ice were launched ahead of the zone of active ice as bergs that floated away in lakes or rivers. In these cases, the bottoms of kettles rest on soft mud or boulder gravel, respectively.

13. I arrange the geography of kettles hierarchically, beginning with a single glaciated fringe. I divide that into five sectors, which are recognized by geologists, but the names are my own. Sectors provide a scientific means of describing the geography but are not to be confused with the

five general descriptive regions defined in the introduction. Each sector is divided into lobes, the common names of which generally follow Mickelson et al., "Late Wisconsin Glacial Record," 4.

14. Some geologists refer to these upside-down Vs as cleavages. The largest kettle on Cape Cod is Long Pond in Brewster and Harwich, covering 743 acres. Deepest are Cliff Pond and Mashpee Pond (Oldale, *Cape Cod*, 64–67). Both the Katama and Mashpee plains resemble modern-day Sprengisandur, an Icelandic outwash plain.

15. Valley kettles are equivalent to what are known as dead-ice sinks in the geological literature. The sand plains flanking them are known in New England as kame terraces, though this name is often a misnomer.

16. Thoreau, *Walden*, 122–23.

17. The major lobes are summarized by Flint, *Glacial Map*: Huron above Detroit; Erie in Pennsylvania, Ohio, and Indiana; Saginaw in Lake Huron; Lake Michigan in Lake Michigan; and Green Bay in eastern Wisconsin. Their surface profiles and rates of motion are given by Clark, "Surface Form." Major moraines that formed during multiple advances, retreats, and still-stands of these lobes bear locally important geographic names. For example, the Huron-Erie Lobe created the Union City, Mississinewa, Salamonie, Wabash, Fort Wayne, and Defiance moraines. The Lake Michigan Lobe created the Shelbyville, Cerro Gordo, West Ridge, Champaign, Urbana, Bloomington, Chatsworth, Marseilles, Manhattan, Valparaiso, and Lake Border moraines, and others.

18. Clark and Walder, "Subglacial Drainage."

19. Wright and Ruhe, "Glaciation of Minnesota and Iowa," identified ten historic phases, four major lobes, and multiple sublobes. The update by Hobbs and Goebel, *Geologic Map of Minnesota*, portrays an even more complex history.

20. The drilling log of boring MGS hole HB-87-1 (*The Itasca Guidebook*, unpublished, for sale at the visitor's center) reports 590 feet of glacial deposits overlying 90 feet of soft material (shale and decomposed rock) above hard Proterozoic bedrock. The basin of Itasca (Wright and Ruhe, "Glaciation of Minnesota and Iowa," 32) lies above a prong of Cretaceous marine shale.

21. The well-known moraines are the Bemis, Altamont, Humboldt, Algona, Fairmont, St. James, Marshall, Antelope, Big Stone, Wahepeton, and Holt. Quote from Keillor, *Lake Wobegon Days*, 13.

22. From Marquart book title, *Horizontal World*. To the north are the Oakes, Kensal and Cooperstown, North Viking, and Martin moraines.

Local relief is typically less than 150 feet. Potholes are most common on moraines.

23. Steinbeck, *Travels with Charley*, 138.
24. Clayton and Moran, "Glacial Process-Form."

## 3. Extinction, Early Humans, and Stabilization

1. Schoolcraft, *Myth of Hiawatha*, 180.
2. Martin and Klein, *Quaternary Extinctions*, provide an excellent overview of the details of the extinction, including a list of fauna. Flannery, *Eternal Frontier*, provides a more popular treatment. Illinois State Museum of Natural History, FAUNMAP, is a searchable database accessible online.
3. Nester and Brown, "Three Dimensional Geometry."
4. Whitman, *Leaves of Grass* ("Song of Myself," stanza 31).
5. Portnoy et al., *Kettle Pond Data Atlas*, 10. Pielou, *After the Ice Age*, 177.
6. Thorson and Webb, "Cedar Swamp." Though the term is not commonly used, a layer of lacustrine mineral sediment is nearly ubiquitous. Palynologists usually refer to its material/texture—sand, silt, clay, marl, volcanic ash, or simply mineral sediment—with the origin as lake sediment assumed.
7. The shape of most lake basins approaches that of an elliptic sinusoid, though small kettles are often more ellipsoidal in shape (Wetzel, *Lake and River Ecosystems*, 36–37).
8. The settling rate of a particle depends primarily on the size of the grain, the density and viscosity of the fluid, and the turbulence as given by Stokes law and the Reynolds equation.
9. Illinois State Museum of Natural History, FAUNMAP.
10. Reported by Fisher, "Taphonomic Analysis." Accepted as evidence by Grayson and Meltzer, "Clovis Hunting."
11. Waters and Stafford, "Redefining the Age of Clovis." Apparently, mastodon were preferred over mammoth because they were smaller, more broadly distributed, later in time, and tended toward solitary behavior. Most archaeologists now believe that human groups were present in the Americas when Clovis technology arose and spread quickly.
12. Flannery, *Eternal Frontier*, 192.
13. Matheus, *Short-Faced Bear*, 135.
14. This discussion does not include capillary motion, which operates independently of gravity with thin tubes and films of water.
15. Such settings receive marine sediment during severe storms, making them valuable archives for hurricane histories at the millennia scale.

16. Since 1931 the Brooklyn Botanical Garden has exhibited the flora and fauna of kettle pools as one of nine important wildflower communities within a hundred miles of New York City.

17. Though the liner isn't necessary to hold the water in these large ponds, it remains as an important barrier to vertical water movements.

18. Depth varies depending on height of the pond surface, which varies with the year. Thoreau reported Walden as being 102 feet deep, but he knew that it was deeper when the pond surface was higher.

19. This section is based on Deevey, "Connecticut Lake Sediments"; Davis, "Holocene Vegetational History"; COHMAP Members, "Climatic Changes"; and especially Webb et al., "Vegetation, Lake Levels, and Climate." The latter summarizes a subset of 851 pollen cores created by more than 100 palynologists. I recommend Jacobson, Webb, and Grimm, "Changing Vegetation Patterns," plate 2, as the most effective published visual representation of vegetation history.

20. Radiocarbon dating is based on the decay of 14C, a carbon isotope produced in the atmosphere by solar processes, incorporated into all organic matter when organisms are alive and decaying spontaneously at a known rate.

21. These are only broad generalizations, as elevation and soil type also control the vegetation of specific sites, with pine/aspen favoring coarser soils at higher elevations, oak/basswood/sugar maple at finer-grained, elevated sites, and prairie at lower elevations. See McAndrews, "Pollen Analysis and Vegetational History," for a regional example.

22. Called the Younger Dryas after a characteristic plant, the interval extended from about 13,000 to 11,650 years ago and was associated with a change in the configuration and hydrology of the large ice sheet to the north. Licciardi et al., "Freshwater Routing."

23. Alkaline water, especially when rich in calcium carbonate, coprecipitates with dissolved phosphorus, removing it as an aquatic nutrient.

24. The cation exchange capacity, or CEC, is a familiar parameter in soil science that expresses the degrees to which the soils can bind and release cations.

25. The word "weed" has no real botanical meaning because a cultural judgment is involved. I use the word as a synonym for what a limnologist would call a macrophyte.

26. Jacobson, Webb, and Grimm, "Changing Vegetation Patterns," plate 2, and Webb et al., "Vegetation, Lake Levels, and Climate," fig. 17.10 (449–50), illustrate this complex story. The thermal battle took place in the zone between two east-west lines: one extending from central Min-

nesota across northern Wisconsin toward Lake Ontario, the other extending from northern Iowa to northwestern Pennsylvania.

27. Webb et al., "Vegetation, Lake Levels, and Climate," figs. 17.14–17.17 (454–57). The actual amount that lake levels fluctuated is not reported in this summary, but was generally fifteen feet or less, usually in the vicinity of several feet. Fritz, "Lacustrine Perspectives," 233–34.

28. Wetzel, *Lake and River Ecosystems*, 274. Aquatic fossils also show little change during the mid-Holocene relative to earlier and later periods.

## 4. NATIVES AND THE LAKE-FOREST ECOSYSTEM

1. Deloria, *God Is Red*, 78, criticizes western approaches as being linear-historic, rather than timeless, and as missing the important intuitive component. To honor both points of view, I treat the ecological aspect of the Indian way of life as timeless, but place it in the context of a time-bound narrative based on rational evidence.

2. Waters and Stafford, "Redefining the Age of Clovis," 1122, place the age of the tradition from 11,050 to 10,800 radiocarbon years before present. The presence of pre-Clovis human occupations in North America is now generally accepted (Grayson and Meltzer, "Clovis Hunting"), but has no bearing on this narrative because there are no accepted sites in the glaciated fringe. I do not suggest that Clovis remains are commonly found at the base of sites, but that they are the common expectation for basal remains.

3. Agriculture was formerly considered a hallmark trait of the Woodland period. However, rudimentary cultivation was present in the late Archaic period, though not in the glaciated fringe.

4. The general perception of long-term lake history is one of steady progressive change and a gradual increase in fertility. Limnologists, however, have demonstrated that this is not the case. Rather, "once organic sedimentation has become established after the initial oligotrophic phase, a type of trophic equilibrium or reasonably stable state occurs" (Wetzel, *Lake and River Ecosystems*, 274).

5. Quote is from Fagan, *Ancient North America*, 102, a useful summary on which the prehistory in this chapter is based.

6. Early Archaic (ten–seven thousand years ago) was transitional from the ice age. Late Archaic (four–two thousand years ago) was transitional to Woodland. This discussion focuses mostly on the middle Archaic, which is most representative of the Holocene ecosystem, and which spanned the warmest period of the present interglaciation. Dates from Kipfer, *Dictionary of Archaeology*, 32.

7. Thomas et al., *Illustrated History*, provide a recent review of the different groups, which were semiautonomous clusters of kindreds. The word "tribe" as used today was introduced by Europeans. Based on genetic profiling using mitochondrial DNA (maternal lineage), most of this group belong to the same broad genetic stock (X Haplogroup), presumed to have arrived from Asia near the end of the Pleistocene.

8. Within the glaciated fringe, the Algonquin language phylum on the Atlantic coast contained the Abenaki, Penobscot, Micmac, and Malecite. In New England they were the Pennacook, Massachusetts, Wampanoag, Narragansett, Niantic, Podunk, Montauk, Wappinger, Mohegan, Pequot, Mahican, and Delaware. In the northern Great Lakes they were the Ojibwe, Ottawa (Odowa), Potowatomi, and Menominee. In the southern Great Lakes they were the Sauk, Fox, Kickapoo, Shawnee, Miami, and Illinois. The Macro Siouan ancestor diversified into the tribal languages of the Great Plains—Mandan, Hidatsa, Omaha, Ponca, Osage, Kansa, Quapaw, and Winnebago—and the Iroquois group of languages: Erie, Huron, Neutral, Tobacco, Iroquois (Mohawk, Oneida, Onondaga, Cayuga, Seneca), Tuscarora, Susquehanna, and Conestaga.

9. Mann, *1491*, 247.

10. This strong generalization is restricted to the central and northern parts of the glaciated fringe. When I suggest the natives were integral members of the food web, I do not imply that they made a "conscious" attempt to live in harmony with nature, for they quickly adapted their ways to accommodate firearms and the export of furs.

11. Ritchie, *Lamoka Lake Site*, 130–32. His subtitle, "The Type Station of the Archaic Algonkin Period in New York," defines it as the standard reference site. Fictile refers to pottery.

12. Reference is to Rousseau's "noble savage," a romanticized view of indigenous peoples.

13. Definition from Stoltman and Baerreis, "Evolution of Human Ecosystems," 252. Archaeology follows Fagan, *Ancient North America*.

14. Sea level had stabilized along the Atlantic shore, creating new habitats of extensive salt marsh, broad sandy beaches, and fertile tidewater estuaries that didn't exist before. Permanent fishing camps developed at seasonal weirs. This more concentrated and more dependable food supply led to rising populations and more sedentary lifestyles, which in turn led to remarkable changes in social structure. From a few early centers near the Louisiana-Arkansas line beginning about five thousand years ago, the Moundbuilders spread far and wide across the eastern seaboard, giving rise to the Adena, Hopewell, and Mississippian cultures. Larger populations led to escalating territorial conflicts and therefore chronic

warfare, hardly an ecosystem in a steady state. Stoltman and Baerreis, "Evolution of Human Ecosystems," differentiate a Cultivating Ecosystem Type and an Agricultural Ecosystem Type, in which crops are supplemental and essential, respectively.

15. "Foraging ecosystems were generally prevalent in the Eastern Woodlands from around 7500 BC to around 2500 BC, with persistence occurring in northern and southern marginal areas up until the time of contact with Europeans." Stoltman and Baerreis, "Evolution of Human Ecosystems," 255.

16. Fagan, *Ancient North America*, 387.

17. Stoltman and Baerreis, "Evolution of Human Ecosystems," 256. Low populations estimated before that time suggest that fishing was essential to the foraging lifestyle.

18. Rostland, *Freshwater Fish and Fishing*, 85 (commas in quote are mine).

19. Morrison, *European Discovery*, 301. Following quotes ibid., 305, 306.

20. Ibid., 374. Morrison points out that they were on a fishing trip, carrying no more than necessary. A sous is a "five-centime piece" according to *Merriam-Webster Dictionary*.

21. This is supported by Verrazano's description of the "Abnaki" of coastal Maine as crude, ill-mannered, barbarous, and clothed in furs. See also Shoemaker, *Strange Likeness*, 3–12.

22. Mason, *Schoolcraft's Expedition*, vii.

23. Covers the period 1610–1791. Another early ethnographer was Lewis Henry Morgan, who published his work on the Iroquois in 1851, significantly after Schoolcraft's fieldwork, but before his publication.

24. Densmore, *Chippewa Customs*, 6. The origin of the tribal name is lost, but Densmore suggests it comes from "to pucker," which refers to the type of moccasin worn, one with a single seam up the middle, rather than around the front, as is the custom today.

25. Ibid., 5.

26. Catton, *Michigan*, 75.

27. Ibid., 115.

28. Schoolcraft, *Expedition to Lake Itasca*, 51.

29. Longfellow, *Complete Poetical Works*. Longfellow wrote, "I pored over Mr. Schoolcraft's writings nearly three years, before I resolved to appropriate something of them to my own use" (Schoolcraft, *Hiawatha*, 15).

30. Trachtenberg, *Shades of Hiawatha*, xii–52.

31. Densmore, *Chippewa Customs*, 8–10 (citing William W. Warren, *Collections of the Minnesota Historical Society*, vol. 5, 78–80, St. Paul: 1885). "Obligate" is a scientific term meaning an organism requiring a habitat.

32. Schoolcraft, *Hiawatha*, 78.

33. Some of these activities were later prohibited for non–Native Americans like me by federal legislation in the early 1970s.

34. Densmore, *Chippewa Customs*, 9.

35. Crawford and Smith, "Paleoethnobotany," table 6.1 (379).

36. Olson, *Runes*, 107–08.

37. LaDuke, *Recovering the Sacred*, 170.

38. Longfellow, *Complete Poetical Works*, 143.

39. Schoolcraft, *Expedition to Lake Itasca*, 34.


## 5. The Fur Trade, the Northwest Passage, and the Source of the Mississippi

1. Quoted phrase is the subtitle of Barbara Tuchman's *A Distant Mirror* (New York: Alfred A. Knopf, 1978). Fagan, *Little Ice Age*, provides a review of this history.

2. Robinson, *History of North Dakota*, 58.

3. Heat Moon, *Blue Highways*, 292.

4. Blegen, *Minnesota*, 77.

5. Smith, "Description of New England," 25.

6. Reviewed by Outwater, *Water*, 11.

7. Merk, *Westward Movement*, 55. 1610 is the earliest possible date for the rendezvous.

8. Catton, *Michigan*, 103.

9. Robinson, *History of North Dakota*, 60.

10. Catton, *Michigan*, 40.

11. Current, *Wisconsin*, 17 (attributed to Felix Keesing, American Philosophical Society, 1939, quoting *Jesuit Relations*).

12. Blegen, *Minnesota*, 164.

13. Morrison, *European Discovery of America*, 415.

14. Grand Portage is an overland trail, mostly on rock and thin till. It is eight miles long, but bypasses the lowest twenty miles of the Pigeon River, a reach that cannot be navigated owing to rapids and waterfalls. Consult Shepard, *Northern Plains*, 105–07, for a review of the setting.

15. White, "Deep Erosion." Most geologists accept that no more than a few tens of meters of vertical exhumation has taken place over most of the shield.

16. Geologists refer to these as underfit streams because the valley in which the river flows was cut by much larger flows in the past.

17. Glacial lakes, many of which were large enough and long-lasting enough to receive proper names, no longer exist. By geological convention, they are distinguished from present-day lakes created by glacial processes by capitalizing the word "Glacial" in front of the lake name. This prefix can be dropped after its first usage.

18. Steinbeck, *Travels with Charley*, 122.

19. Sandburg, *Complete Poems* ("Prairie"), 79.

20. In common geological use, this French word is normally reserved for the notched meltwater channels of the High Plains (Pielou, *After the Ice Age*, 25), though coulees are equivalent to the less-dramatic flumes of New England.

21. John Mitchell's early but undated *Map of the British and French Dominions in North America*, which influenced the boundary decision between Britain and the United States in 1783, assumed that the headwaters were somewhere to the west of Minnesota near the fiftieth parallel. Map and notation reproduced in Fife and Freeman, *Old Maps*, 183–84.

22. Except for the last few miles, along the Pigeon River, the international boundary follows a chain of lakes.

23. The north-south King-Hawksbury line partitions Minnesota from Canada near Lake of the Woods. Its southern extension to the latitude of the source of the Mississippi River was ready for signing the same year the Louisiana Purchase was being negotiated. When it was realized that the Mississippi source likely lay east of the King-Hawksbury line, the decision was made in 1818 to settle the border at the forty-ninth parallel (Lass, *Minnesota*, 55–56).

24. Blegen, *Minnesota*, 112.

25. The pigeon he refers to is likely the extinct passenger pigeon. The unnamed men on the Schoolcraft expedition referred to these shaking savannahs as "têtes des femmes," suggesting that they were French.

26. Schoolcraft, *Expedition to Lake Itasca*, 35.

27. Elk were much more common during the exploration phase prior to farming and logging.

28. Mason, *Schoolcraft's Expedition*, xxiii.

29. Schoolcraft, *Expedition to Lake Itasca*, 51, was particularly amazed at how the natives were seemingly incapable of Christianity, ruefully commenting that "it should not excite surprise that the people themselves are, to so great a degree, mentally the same in 1832, that they were on the arrival of the French in the St. Lawrence in 1532 [actually 1534]."

30. Spencer et al., *Native Americans*, 365.

31. Brown, *Wounded Knee*, 37–65.

32. Trachtenberg, *Shades of Hiawatha*, xvii.
33. LaDuke, *All Our Relations*, 116.

## 6. KETTLES AND EARLY AMERICA

1. Thomas, *Cranberry Harvest*, 31, reviews the early history of resources from kettles based on the work of Henry S. Griffith.
2. Donahue, *Great Meadow*, xv.
3. General discussion of cranberry history based on Thomas, *Cranberry Harvest*, 9–34.
4. Ibid., 13.
5. Thoreau, *Walden*, 199.
6. Richardson, *Emerson*, 139.
7. Conn, *Literature in America*, 169.
8. Donahue, *Great Meadow*, 27–31.
9. Blanchard, *Margaret Fuller*, 86. Zochert, *Walking in America*, 63. The water in Fresh Pond is supplemented by water imported from Weston Reservoir.
10. Thoreau, *Walden*, 125. Subsequent quotes are ibid., 58, 58, 125, 127, 127.
11. Ibid., 130.
12. Ibid., 129, 188. Next quote ibid., 117.
13. Trachtenberg, *Shades of Hiawatha*, xii.
14. Schoolcraft, *Hiawatha*, 7.
15. Bolles, *Ice Finders*, provides an excellent summary of the history of the concept and Agassiz's role.
16. Thorson and Brick, "Stratified Drift."
17. Totten and White, "North American Craton," 127. This organization went on to become the American Association for the Advancement of Science, today the premier scientific organization within the United States. This demonstrates the importance of geology in general and glacial geology in particular during the emergence of American science.
18. Maynard, *Walden Pond, a History*, 203. Following quote ibid., 203.
19. Karl Koteff completed the story in the 1960s when he published the U.S. Geological Survey's *Surficial Geological Map of the Concord 7½ Minute Quadrangle*.
20. Blegen, *Minnesota*, 68.
21. Lewis, *Main Street*, 506.
22. Totten and White, "North American Craton."
23. Whittlesley, "Drift and Alluvium." Louis Agassiz doesn't merit that distinction because he spent most of his career studying paleontology and

zoology. Flint, "Glacial Geology," reviews the early history of the discipline.

24. Totten and White, "North American Craton."

25. Charlesworth, *Quaternary Era*, 44. Black, *Scientific Reserve of Wisconsin*, 31.

26. Chamberlain's final compilation was based on work by W. Upham (1877) on Long Island and Cape Cod, H. C. Lewis (1884) in Pennsylvania, and G. H. Cook and J. C. Smock (1877) in New Jersey. Refer to Charlesworth, *Quaternary Era*, 417.

27. Charlesworth, *Quaternary Era*, 416.

28. Blegen, *Minnesota*, 190.

29. Lass, *Minnesota*, 115.

30. Blegen, *Minnesota*, 305.

31. Quote and paraphrased text from Lass, *Minnesota*, 15.

32. Atwood, *Physiographic Provinces*, plate facing p. 10.

33. Blegen, *Minnesota*, 307.

34. Catton, *Michigan*, 161.

35. Defebaugh, *Lumber Industry*, 1–20.

36. Hemingway, *Complete Short Stories* ("Big Two Hearted River"), 213.

37. LaDuke, *All Our Relations*, 4.

38. Blegen, *Minnesota*, 331.

39. National Geographic Society, *Historical Atlas*, 139, reports this as 7,140,000,000 cubic feet.

40. Lass, *Minnesota*, 149.

41. Blegen, *Minnesota*, 330.

42. In legend, Babe's blue color is sometimes attributed to the "year of blue snow," a phenomenon that can happen if thick early snowfalls have a chance to recrystallize and don't melt until after calving season. In other accounts, the blue comes from the veinous color of hypothermic skin after Babe was pulled from an ice-covered lake.

43. Shay, *Itasca Bison Kill Site*.

44. Paleontologist R. Dale Guthrie spent several years examining and evaluating this steppe bison (*Bison priscus*), which was even larger than *Bison occidentalis*. He concluded that it had been killed by lions before being buried by a cold mudflow more than 35,000 years ago. Lions existed in kettle lake country prior to the extinction about 13,000 years ago.

45. Winkler, "Changes at Walden Pond."

46. The Menominee Range in Michigan's Upper Peninsula; the Gogebic Range in Wisconsin; and the Cuyuna, Vermillion, and Mesabi ranges in northeastern Minnesota.

47. Blegen, *Minnesota*, 361.

48. Catton, *Michigan*, provides an overview of Michigan's mining industry.

49. Mosher, *North Country*, 85.

50. Ibid., 84. Quote attributed to Harlan H. Hatcher (*The Great Lakes*, New York: Oxford University Press, 1944).

51. Ibid., 17.

## 7. Family Lake Culture

1. Cronon, *Nature's Metropolis*, 380. "The pastoral retreat in its mythic form is a story in which someone becomes oppressed by the dehumanized ugliness of urban life and so seeks escape in a middle landscape that is halfway between the wild and the urban."

2. Newport, Rhode Island, was the premier resort destination for wealthy easterners during the Gilded Age. Web site from the Geneva Lakes Area Chamber of Commerce, http://www.lakegenevawi.com/history.htm (copyright 1999–2007).

3. Maynard, *Walden Pond, a History*, 116.

4. For subsistence, Thoreau took many of his meals at the family house in Concord, walking back to Walden with leftovers in his pocket. Though he proclaimed self-reliance as a virtue, he was financially and emotionally dependent on his mother and sister and the family of his patron, Ralph Waldo Emerson, especially his wife, Lydia. William Ellery Channing also envisioned Thoreau's house as a "wooden inkstand," not unlike the writing shacks of other famous authors, little more than a quiet place to read and write while being free from the distractions of town.

5. Quote from Horace Mann Jr., Thoreau's seventeen-year-old traveling companion from a letter to his mother. Thoreau's friends, including the boy's father, Horace Mann, were not able to accompany him.

6. Halprin and Kattelle, *Lake Boon*, 7.

7. Lingeman, *Sinclair Lewis*, 1–75.

8. Lewis, *Main Street*, 515, 412.

9. Ibid., 88–89.

10. Löfgren, *On Holiday*, 280.

11. Lynd and Lynd, *Middletown*, 259.

12. North, *Rascal*, 79.

13. Baker, *Ernest Hemingway*, 3. The Hemingway camp lay near the south end of Walloon Lake in northwestern Michigan. Lakes near the edge of the peninsula tend to be rockier; those in the interior are usually kettles.

14. Löfgren, *On Holiday*, 132, 135, 137.

15. White, "Once More," 182.
16. Lake Koshkoning is one of a chain of lakes lying within a prominent kettle moraine (Hadley and Pelham, *Glacial Deposits*) comprised mostly of partially collapsed sand and gravel (Soller, *Quaternary Sediments*).
17. Steinbeck, *Travels with Charley*, 97; ibid., 95.
18. Hutchinson, "Prospect Before Us," 689.
19. Cronon, *Nature's Metropolis*, 378.
20. Historian of the West Frederick Jackson Turner wrote in his famous 1893 paper, *The Significance of the Frontier in American History*, that "in the crucible of the frontier, the immigrants were Americanized, liberated, and fused into a mixed race."
21. Minnesota State Department of Natural Resources, "Minnesota Facts."
22. Sinclair Lewis began his book about Minnesota when living in Chatham, Cape Cod, and within easy reach of a dozen stunningly beautiful kettles. As with Thoreau, he mostly ignored them.
23. Whitteman and Webster, *Water Escapes*. Only five lakes are mentioned. All are large bedrock lakes.
24. Lake Baikal in Russia contains more water than all the Great Lakes combined, but is a narrow sliver in terms of surface area. The Caspian Sea and the Black Sea are larger but too salty to be considered freshwater. Together, the Great Lakes hold 24,620 cubic kilometers of water.
25. Castillo, "Illinois," 126.
26. Thoreau, *Walden*, 118.
27. Frazier, *Fish's Eye*, 74.
28. Olson, "Northern Lights," 433.
29. Connors, "Biography of Walt Whitman," 12.
30. This is clear from a reconnaissance through the Federal Writers Project documents for various kettle states, 1936–1940, Library of Congress. Lakes were places and environments, not literary devices.
31. White, "Walden—1954," 360.
32. Many of these writers followed Thoreau's example by isolating themselves in a rural setting within easy reach of society, especially in Connecticut, where Joseph Wood Krutch bought a farmhouse in Redding, Edwin Way Teale created a miniature Walden Pond on his farm in Hampton, and Michael Pollan hoed his own beans in the Litchfield Hills.
33. The Paul Harvey news program *The Rest of the Story* may be a better-known radio show, but it is not a variety show.
34. Despite Keillor's tongue-in-cheek assertion that the name has Ojibwe

origins, Wobegon must be a shortened version of *woebegone*, a word meaning woeful or in a sorry state, meaning that woe is present rather than gone. There is no entry for the name or anything similar to the name in Baraga's *Dictionary of the Ojibway Language* (422).

35. Keillor, *In Search of Lake Wobegon*, 17. The 1985 novel *Lake Wobegon Days* describes the lake in some detail and features it on the cover.

## 8. How Lakes Work

1. The word "oceanography" is a historic misnomer, a holdover from the age of sextant and sail, when the main objectives were to keep track of geographic position and to differentiate the seven seas. The term "oceanology" better describes what takes place today. The founder of limnology, François-Alphonse Forel, made a conscious choice to avoid the term "limnography." G. E. Hutchinson ("Prospect Before Us," 683–84) suggests that the main difference between oceanography and limnology is that the latter cannot ignore the boundary of the water mass.

2. Organization of limnology at the national level actually began in 1925 with the formation of the Committee on Aquaculture of the American Association for the Advancement of Science, but limnology as a science was still embedded in another field, zoology.

3. Section condensed from Frey, *Limnology*. Limnology was organized within Europe by 1922, when the Swedish scientist Einar Naumann and August Thienemann formed the Societas Internationalis Limnologiae.

4. Wetzel, *Lake and River Ecosystems*, 4.

5. Wetzel, *Limnology*, 134.

6. Hagen, *Entangled Bank*, provides a detailed history.

7. Both Deevey, "Connecticut Lake Sediments," and Davis, "Phytogeography," are built on earlier work in Sweden by Jakob Ljungqvist and Lennart von Post.

8. Thoreau's contributions to limnology have been largely overlooked because the main contributions of *Walden* are in literature, philosophy, and social criticism. According to the pioneering limnologists John Brooks and Edward Deevey Jr. ("New England," 118–19), "two chapters [of Walden] are devoted to original scientific observation of high quality."

9. Thoreau, *Walden*, 189. Following quote ibid., 131.

10. In the ocean, water density is controlled more by salinity than by temperature. The reverse is true for lakes. The reduced density contrasts in lakes cause more sluggish currents.

11. Limnologists refer to the well-mixed, generally oxygenated, and warm upper layer as the epilimnion and to the more stagnant, colder, and often less-oxygenated deeper layer as the hypolimnion. They also distinguish a third layer, the metalimnion, in which the water is stationary but the temperature decreases with depth.

12. Lakes that mix twice a year are referred to as dimictic lakes. The northern half of the glaciated fringe and the zone of thick winter ice are nearly coincident, meaning that kettles are usually dimictic.

13. Dingman, *Physical Hydrology*.

14. Thoreau, *Walden*, 122.

15. This so-called residence time, typically ranging from several months to several years for kettles, is calculated by dividing the volume of the lake by the rate of inflow or outflow.

16. The degree to which groundwater impacts lake chemistry is proportional to the amount of glacial sediments relative to bedrock (Winter and Woo, "Hydrology of Lakes"). Kettles are effectively 100 percent glacial sediment, therefore groundwater inflows control the chemistry.

17. Salts are generally inorganic compounds that freely dissolve into anions and cations and recombine to form electrically neutral compounds in water.

18. Swanson et al., *Chemical Characteristics of Prairie Lakes*, no pagination.

19. There are exceptions, most notably bacteria that create organic matter geothermally or through the iron-sulfur reactions.

20. Aerobic (with oxygen) respiration depends on adenosine 5-triphosphate, the main molecular means of energy transfer within cells.

21. Molecular nitrogen ($N_2$) must be converted into ammonium ($NH_4$), nitrite ($NO_2^-$), or nitrate ($NO_3^-$) to be biologically useful. In lake systems, nitrogen fixation takes place mostly by bacteria within the oxygen-poor conditions of the lake bottom, where anaerobes (especially blue-green algae) thrive. It also takes place within the roots of shrubs, particularly alders, where symbiotic bacteria live. Biologically available nitrogen also arrives via air and water pollution, from surface runoff, and from groundwater after drainage through soils.

22. The productivity of a lake is measured by the amount of organic matter produced in a given period of time, usually a year. More specifically, it is the amount of carbon "fixed" by photosynthesis from aquatic plants. The most reliable single predictor of a lake's productivity is its mean depth, with deeper lakes being less productive.

23. Of the three nutrients, phosphorus is usually most limiting, followed by nitrogen, and only rarely by carbon. Aquatic organic matter typically has

a 1:7:40 ratio of P:N:C. "Phosphorus can theoretically generate 500 times its weight in living algae, nitrogen 71, and carbon 12 (Wetzel, *Lake and River Ecosystems*, 275).

24. These purely physical relationships run opposite those associated with biological activity, in which cold temperature and high pressure are usually limiting. Though cold water can hold more oxygen, there is often less oxygen dissolved in it because cold water reduces the rate of photosynthesis. And though deep water is under higher pressure and thus can hold more oxygen, it is further removed from its two main sources, photosynthesis and the atmosphere.

25. The scientific terms for ultralean, lean, fertile, and overfed are dystrophic, oligotrophic, mesotrophic, and eutrophic, respectively.

26. Subsurface geophysical imaging techniques such as seismic reflection and ground-penetrating radar allow scientists to fill in the blanks between borings and core locations.

27. Thoreau, *Walden*, 188.

28. Ice at its freezing point has a dimensionless linear coefficient of thermal expansion of $51 \times 10^{-6}$ length/length. Ice on a mile-wide lake will expand roughly five feet if it warms up forty degrees Fahrenheit.

29. These thermal offsets pose a problem for lake health because the warmer water temperatures of late summer increase the microbial consumption of oxygen at a time when the water can hold less of it.

## 9. Habitats, Flora, and Fauna

1. Muir, "Boyhood," 60.

2. The following discussion of the kettle pond ecosystem is highly selective, anecdotal (based largely on my own experience), emphasizes familiar organisms, and avoids taxonomic names. It was also informed by Thompson, Coldrey, and Bernard (*The Pond*), Mitsch and Gosselink (*Wetlands*), Wetzel (*Lake and River Ecosystems*, chapters 15–22), and Burgis and Morris (*Natural History*), all of which provide a review of the biological systematics.

3. Water is 775 times as heavy as an equivalent volume of air at standard temperature and pressure (density). Its resistance to fluid flow (viscosity) is approximately 100 times greater.

4. Muir, *Reflections in Bullough's Pond*, 27.

5. Thompson et al., *The Pond*.

6. All wetlands must meet three criteria: Water must be present at saturation for much of the year, either within the root zone or above it. The

soils must show evidence of poor oxygenation, either an abundance of organic matter or diagnostic blotches of bluish gray or greenish colors. Finally, the vegetation, if present, must be specifically adapted to living under wet conditions (Mitsch and Gosselink, *Wetlands*, 25–34).

7. Historically (Schoolcraft, *Expedition to Lake Itasca*, 4), a wetland associated with a kettle was referred to as a morass, bog, swamp, or savannah in the heartland and as a marsh, bog, or fen in the east (Johnson, *Bogs*).

8. Ibid., 7.

## 10. Loving Lakes Too Much

1. The Red Lake, Cass Lake, Leech Lake, and White Earth reservations lie to the north, east, south, and west of Bemidji, respectively.

2. This form of secondary treatment, developed in the 1950s, uses bacterial slime to clean the water. Phosphorus was not chemically treated until the late 1970s.

3. Hundreds of news stories in print and electronic media featured gray green water flowing beneath the bridge collapse on Interstate 35W during the summer of 2007. Water-quality data indicated total phosphorus concentrations were well in excess of the desired amount.

4. Schoolcraft, *Expedition to Lake Itasca*, 254. Following quote ibid., 204.

5. Lake Plantagenet has many attributes that make a lake prone to nutrient enrichment. Being broad and shallow, much of the water lies within the productive photic zone and is warmed by solar radiation. The lake flushes slowly, due to the large volume relative to the normal trickle of its inlet and outlet discharges. During stronger flows in spring, the inlet carries nutrient pollution because it drains a fairly large semiagricultural watershed that is also being managed for forestry. Much of the shoreline was developed for recreational cottages at a time when the filling of wetlands was viewed as a shoreline improvement and when wastewater disposal methods were primitive.

6. Hawthorne, "Hobnobbing," 63.

7. Winkler, "Changes at Walden Pond."

8. Maynard, *Walden Pond, a History*, 238.

9. Ibid., 326.

10. Colman and Waldron, *Walden Pond*.

11. Oldale, *Cape Cod*, 170.

12. Citro, *Weird New England*, 21.

13. Finlayson and D'Cruz, "Inland Water Systems."

14. Johansen, *Dirty Dozen*.

15. Portnoy et al., *Kettle Pond Data Atlas*, 43–48.

16. Harris and Steeves, *Cape Cod Aquifer*.
17. Ibid., 41.
18. The technical synonym for "overfeeding" is "eutrophication," defined as an increase in the biological productivity of the system above some baseline condition as measured by the amount of carbon being fixed.
19. The threshold value for dissolved oxygen (DO) is about 0.5 mg/L.
20. Though common in folklore, the scientific basis for these lights defies easy explanation. The rapid oxidation (combustion) of methane is the likely explanation in most cases. Burgis and Morris, *Natural History*, 43.
21. This happens when nutrient is depleted at higher levels in a static summer situation, but is present at greater depths.
22. This is especially well documented in one of the most well-studied kettles, Mirror Lake in New Hampshire. There paired watersheds draining into the lake were experimentally deforested and the nutrient flux monitored. Refer to U.S. Department of Agriculture (Forest Service), "Hubbard Brook."
23. Wetzel (*Limnology*, 288) claimed that the industry "sentenced thousands of lakes to a status of semipermanent eutrophication for the sake of making more money."
24. Nitrogen is more challenging to regulate because its behavior is so variable. It arrives with the rain or snow falling on the lake surface. Some of this is from automobile exhaust. The rest comes from dust, pollen, natural and industrial organic gases, and aerosols from wildfires and fossil-fuel generating plants. Nearly a third of the nitrogen arriving in many lakes comes from groundwater, either through direct seepage or indirectly through streams fed by groundwater. The remaining sources are nitrogen fixation by wetland plants, rural runoff, and human wastes. The majority of nitrogen loss occurs by sedimentation, meaning the thickening of muck.
25. Quote from Carol Peterson of the U.S. Environmental Protection Agency, who presented her paper on June 9, 2007, at the New England Conference of the North American Lake Management Association, held in Storrs, Connecticut. Accessed from EPA Web site, *National Lakes Assessment*.
26. Ken Wagner, past president of the North American Lake Management Society (June 9, 2007), and Gene Likens, a distinguished ecologist/limnologist from the Institute of Ecosystem Studies in New York (October 2007).

## 11. LAKE FUTURES

1. Harrison, "Michigan," 198–99.
2. Text posted on the Web site http://www.lakeshoredreams.com/ in August 2007. Eric Canfield was listed as the author.

3. Policies for renovation are far more lenient than for new construction, in some cases allowing nearly a 100 percent makeover, provided that at least one remnant of the original structure is retained.

4. Louv, *Last Child*, 16–19.

5. A recent trend is to weigh down the back end of a powerful boat to create a large wake, like moguls on a ski slope, over which a water-skier can launch himself into the air. Such large waves in bays damage shoreline plant and animal communities.

6. Meehl et al., "2007 Global Climate Projections"; Christiansen and Hewitson, "Regional Climate Projections"; and National Assessment Synthesis Team, *Climate Change Impacts*.

7. Projection for 2071–2100 is based on the average weather for the interval 1961–1990 (Christiansen and Hewitson, "Regional Climate Projections").

8. Ibid., 889 (fig. 11.11). Projected regional increases are for two to six degrees Celsius in the central United States and two to four degrees Celsius in the east.

9. The water balance includes all gains (precipitation, inlet streams, groundwater) and losses (evaporation, transpiration, outlets, groundwater).

10. Christiansen and Hewitson, "Regional Climate Projections," 885.

11. Meehl et al., "2007 Global Climate Projections," 750.

12. National Assessment Synthesis Team, *Climate Change Impacts*, 173. A killer heat wave is defined as three consecutive days when daytime highs reach one hundred degrees Fahrenheit (thirty-eight degrees Celsius) and when the nighttime low temperature remains above eighty degrees Fahrenheit (twenty-seven degrees Celsius).

13. Jensen et al., "Ice Phenology."

14. The corollary of being a great place to "get away from it all" is to be in a place that is practically unknown beyond the local community.

15. Winter and Woo, "Hydrology of Lakes," review the differences. Lake basins in rock are much more likely to be linked to the stream network, and their groundwater inflows do not alter lake chemistry nearly as much. Refer to Dingman, *Physical Hydrology*, for thorough treatment.

16. In sequence, the main kettles for the Mississippi overflow system are Elk, Itasca, Carr, Irving, Bemidji, Wolf, Andrusia, Cass, and Winnibigoshish lakes. Only after the last lake does the flow begin to resemble a true river.

17. Its National Water Quality Assessment Program (NAWQA) is explicitly focused on the trinity of rivers, streams, and groundwater. U.S. Geological Survey, National Water-Quality Assessment (NAWQA) Program.

18. Lakes are not listed on the twenty-five featured links on the home page for the EPA's Office of Water, accessed August 2008.

19. For a review of the EPA history relative to lakes, go to http://www.epa.gov/nps/Section319/qa.html.

20. Of the thirteen components on the Web page, none are identified as lake related. Under the component titled "Aquatic Resources Monitoring" are five separate headings, one of which is for lakes (others are Estuaries, Streams, Great Rivers, and Wetlands).

21. Leakey and Lewin, *People of the Lake*.

22. Erdich, *Ojibwe Country*, 77.

# Index

Note: page references in italics refer to illustrations. Those followed by n refer to notes with note number.

## A NOTE ON THE AUTHOR

ROBERT M. THORSON is a professor of geology at the University of Connecticut, where he holds a joint appointment in the Department of Ecology & Evolutionary Biology and the Department of Anthropology. He writes a regular op-ed column for the *Hartford Courant*. He grew up in the upper Midwest, where he developed a lifelong attachment to kettle lakes. Thorson is the author of *Stone by Stone*, which was awarded the 2003 Connecticut Book Award for nonfiction, *Exploring Stone Walls*, and *Stone Wall Secrets*. His work on stone-wall preservation has been recognized in articles in the *New York Times*, the *Washington Post*, the *Boston Globe*, in many magazines, and on National Public Radio. He lives in New England.